REGIONAL ECONOMIC ANALYSIS OF INNOVATION AND INCUBATION

to Yvonne and Mirjam

Regional Economic Analysis of Innovation and Incubation

E. J. DAVELAAR
Free University
Amsterdam

Avebury

Aldershot · Brookfield USA · Hong Kong · Singapore · Sydney

Published by
Avebury
Gower Publishing Company Limited
Gower House
Croft Road
Aldershot
Hants GU11 3HR
England

Gower Publishing Company
Old Post Road
Brookfield
Vermont 05036
USA

ISBN 1 85628 139 6

Printed in Great Britain by
Billing & Sons Ltd, Worcester

Contents

Chapter 2:

Innovations in Spatial Analysis

Chapter 3:

New Technological 'Trajectories' and Spatial Dynamics

Chapter 4:

A Theoretical Model for a 'Dynamic Incubation' Framework

Part B   EMPIRICAL INVESTIGATIONS

Chapter 5:

Structural Changes in the Amsterdam Area

Chapter 6:

Regional Variations in Innovative Performance

Chapter 7:

Spatial Variations in Innovation Potential

Chapter 8:

Generation of Product and Process Innovations

Chapter 9:

Spatial Dispersion of Producer Services in the Netherlands

Chapter 10:

Conclusion

Subject Index 319

References

Preface

In recent years, technological change is increasingly recognized to be one of the prime tools of capitalist competition and economic growth. In accordance with this general 'upswing' in the interest of the role of technological change, the generation and diffusion of innovations in a spatial context is, nowadays, a popular theme in both regional and urban research. In the present study I will focus on the role of innovative activities in generating spatial dynamics.

On the basis of a 'swarming process' through the more recent literature, in the first chapters of this study two central research questions will be identified. The first basic research issue selected in this way, is concerned with the impact of both intra-firm and spatial variables on the innovativeness of individual firms. The second research issue deals with the development - and empirical testing - of a theoretical framework concerning new life cycles of industries and technologies in a spatial context. The analysis of both research issues will be elaborated upon in the present study.

x

The study presented here has been carried out at the Free University in Amsterdam and was financially supported by the Netherlands Organization for Scientific Research (NWO, project number 450-230-004). The research 'trajectory' leading to the present study has been characterized by the well-known 'cycles of inspiration'. Especially in the final phases of finishing this study, I sometimes felt like the prodigal son and I owe a great debt of gratitude to several persons who have supported me in one way or another.

The support of Professor Peter Nijkamp has been indispensable throughout the study. Especially in later phases of writing, I owe him many thanks for keeping me, like a careful father, on the right track. In this context, I also wish to thank Professor Zegveld for making several useful remarks during the final stage of finishing this study. I am grateful to dr. Alfred Kleinknecht for his support in collecting and supplying the necessary data.

Furthermore, thanks are also due to Winny Browne who carefully read and improved my 'English' manuscript. In a similar vein I am indebted to dr. Annemarie Rima who was always prepared to maintain an 'on-line' connection when I encountered hardware or software problems. The study has benefited from stimulating discussions with my colleagues. In this respect, I am especially deeply indebted to dr. Piet Rietveld and drs. Frits Soeteman for their useful suggestions and moral support at times when 'destructiveness' dominated 'creativeness'.

Last but not least my admiration goes to my wife Yvonne, who by means of her patience and dedication demonstrated that even writing a book can coincide with intensive nuptial ties, and my daughter Mirjam who bravely carried the burden of having a 'part-time' father in the first year of her life.

Evert Jan Davelaar
Voorthuizen, April 1989

PART A
THEORETICAL
REFLECTIONS

1 Technological change and economic transformation

1.1 INTRODUCTION

In recent years a renewed interest in the role of technological change as an engine of economic growth, and vice versa, has surfaced. After a period of relatively little interest, in which technological change was mainly considered as 'manna from heaven', times have changed. This renewed interest can partly be ascribed to the decline of economic growth in the seventies. In this period the standard neo-classical growth model, which appeared to offer a more or less adequate description of the 'stable' economic growth experienced in the fifties and sixties, lost much of its appeal and explanatory power. Accordingly, attention shifted to those models and concepts which could explain long-term fluctuating patterns more adequately.

The current 'upswing' in the interest in the *Kondratiev cycle*, as well as in its underlying causal mechanisms, has to be understood in this context. Although the mere *existence* and *cyclicity* of the Kondratiev cycle still remains a matter of scientific dispute, one observes some common agreement regarding the existence of 'structural adjustment periods'. During such periods economic growth based on existing industries and markets stagnates, whilst future economic growth largely depends on the generation of *new markets and industries* (cf., van der Zwan 1979, Mensch 1981a, 1981b, Freeman et al 1982, Rothwell and Zegveld 1985). The current popularity of Schumpeterian economics-especially concerning the role of innovative entrepreneurs as the 'engines' of capitalist economic growth - can be understood in this light.

Consequently, innovations are increasingly considered to be one of the prime tools to (re)generate economic growth. Innovative firms have to seek new markets in order to fight against the above-mentioned market saturation of 'old line industries' and rigid production structures. So, besides price competition based on existing products - as stressed in the neo-classical theory - competition based on *product innovation* is increasingly considered fundamental to economic growth.

In this respect, it is increasingly acknowledged that technology cannot be considered as an exogenous 'deus ex machina' which raises the capital and/or labour productivity automatically and permanently. In this context, firms purposely strive for the generation of innovations, for example by means of R&D efforts. So it is increasingly recognized that innovative activities and economic performance are positively correlated, although the *direction of causality* may sometimes appear ambiguous, as will become clear later on in the present chapter.

In this chapter our primary purpose will be to select, discuss, and integrate *several concepts* from the *recent innovation literature*. We do not aim at presenting an exhaustive picture however: the main purpose of this chapter lies in sketching an 'integrating framework' that will serve as a basis for our analysis in subsequent chapters. In this context, we will take a long-term perspective, and also approach the issue of the interdependence between innovative activities and structural economic changes accordingly. In the following chapters this framework will be linked to the spatial dimension of innovative behaviour.

For this purpose, we will briefly discuss a set of relevant 'topics' concerning the relationship between economic performance and innovations in the next section. Then in section 1.3, the empirical dimension of this relationship will be considered. These sections will will constitute the basis for our sketch of the 'integrating framework' in subsequent sections. Consequently, in section 1.4 we will start with a discussion of some abstract but nevertheless useful concepts like new 'technology systems', technological 'paradigms' and technological 'trajectories'. Next, in section 1.5 we will discuss the interdependencies between new life cycles of industries and technologies. Finally, in section 1.6 the most important conclusions of this chapter will be recapitulated.

1.2 INNOVATIONS AND ECONOMIC PERFORMANCE

As mentioned in the introduction to this chapter, innovations are increasingly considered to be of prime importance to the competitive performance of firms, sectors, regions and countries. By means of *product innovations* - which will for the time being be conceived of as new or modified (i.e., improved) products or services - the innovator may gain a competitive edge over his competitors when this innovation meets consumer preferences and tastes better than its substitutes. Consequently, competition based on product innovation may be an effective way for increasing profits and sales.

Process innovations on the other hand - which will be considered to be new improved process technologies concerning existing products or services - may be an important means for increasing productivity and, consequently price competitiveness. Especially these innovations have been dealt with in the well-known neo-classical models of perfect competition, largely ignoring the issue of product innovations. In this respect, especially Schumpeter can be credited for being one of the first economists also recognizing the importance of *product innovations* to economic competition. As regards the generation of innovations two 'Schumpeterian' models are often suggested dealing with the degree of autonomy of the technology sector (cf., Freeman et al 1982, Winter 1984).

In the Schumpeter I model the role of innovative entrepreneurs is considered to be fundamental to the process of 'creative destruction'. This aspect emerges by the generation of new 'combinations' or (product) innovations (cf., Schumpeter 1934), superseding previous 'combinations'. This process of 'creative destruction' is considered to an important capitalist engine of economic growth.

Especially *new* innovative entrepreneurs entering upon the scene are considered to perform the 'task' of carrying through these new combinations: 'new combinations are, as a rule, embodied in new firms which generally do not arise out of the old ones but ,start producing beside them' (Schumpeter 1934, pp.66). In the Schumpeter I model, these new combinations stem especially from progress in scientific research. For these entrepreneurs the 'carrot' for introducing such combinations is the prospect of a temporary monopolistic rent.

Other potential entrepreneurs, observing the success of some of the first innovator(s), will soon try to catch up by means of *imitation and further improvements* of these new combinations. So following the initial innovators, a kind of 'bandwagon' or 'swarming' process will soon set in. A phase of economic expansion will commence as more entrepreneurs emulate the initial innovators. During this expansionary phase, the 'carrots' of excessive profits will become smaller as more entrepreneurs enter upon the scene and the new markets approach *saturation*. This decreasing (expected) profitability, during the 'life cycle' of such new industries, will serve as a limit to the booming phase. So, in

> Schumpeter's framework it is disequilibrium, dynamic competition (in the sense of 'imperfect' competition) among entrepreneurs, primarily in terms of industrial innovation which forms the basis of economic development (Freeman et al 1982, pp.31).

Freeman et al (1982) summarize this Schumpeterian model of 'entrepreneurial innovation' in the following way:

Figure 1.1 Schumpeter's Model of 'Entrepreneurial Innovation'

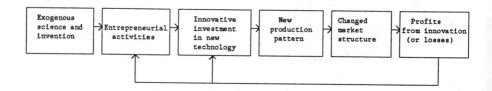

Source: Freeman et al 1982, pp.39

This Schumpeter I model is again quite popular nowadays. Especially empirical studies like Birch's analysis (1979), have credited this model version. This study demonstrated that the greater part of the new jobs created in the 1970s in the US could be ascribed to small, especially non-manufacturing firms, most of which existed only a few years. In this context, Naisbitt (1984, pp.7), in discussing the current transition from an industrial to an information economy (see also Chapter 9 of this study), states: 'The transition times between economies are times when entrepreneurship booms. We are now in such a period.' Also from a policy point of view a (re)interest in the role of new innovative firms can be observed (cf., Rothwell and Zegveld, 1985).

In his later writings, Schumpeter however, stressed the fact that technology became increasingly integrated (i.e., endogenous to) the economic system (see Schumpeter 1943). Schumpeter derived this thesis from the observation that in the twentieth century especially large firms set up industrial research laboratories. In this way technology became institutionalized within *existing* firms. The basic difference with the Schumpeter I model lies in the

> incorporation of endogenous scientific and technical activities
> conducted by large firms. In Schumpeter II therefore there is a
> strong positive feedback loop from successful innovation to
> increased R and D activities setting up a 'virtuous' self-
> reinforcing circle leading to renewed impulses to increased market
> concentration (Freeman et al 1982, pp.41).

The positive feedback between innovative activities and firm size in the Schumpeter II model, tends to weaken the entrepreneurial function, however. Freeman et al (1982) summarize 'Schumpeter's model of 'large-firm managed innovation' as follows:

Figure 1.2 Schumpeter's Model of Large-Firm Managed Innovation

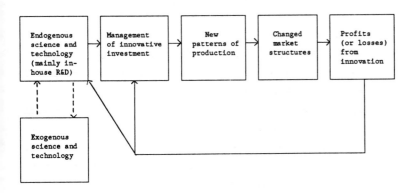

Source: Freeman et al 1982, pp.40

Essentially, both 'Schumpeterian' models of innovation have been at the heart of many subsequent discussions whether *small firms* or *large firms* should be considered as the main generators or adopters of new 'combinations' and the concomitant economic growth. As stated above, the Schumpeter I version is regaining considerable interest nowadays. In the next sections however, we will reconsider this discussion and argue that the relative importance of both models may, inter alia, be related to the *different phases* of new technologies.

Another issue in the debate on the sources of technological change, refers to the so-called *'technology-push'* versus *'demand-pull'* discussion. The essence of this argument is whether innovations are 'science' driven, or generated because of a demand-pull effect. Especially Schmookler (1966) has stressed the latter effect. In a study based on patent statistics - in particular related to the capital goods sector - and investments, Schmookler concluded that *changes in demand* appeared to *precede patent cycles*:

> When time series of investment (or capital goods output) and the number of capital goods inventions are compared for a single industry, both the long-term trend and the long swings exhibit great similarities, with the notable difference that lower turning points in major cycles or long swings generally occur in capital goods sales before they do so in capital goods patents (Schmookler 1966, pp.204).

From this analysis Schmookler concludes: 'the production of inventions, and much other technological knowledge, whether routinized or not, ..., is in most instances as much an economic activity as is the production of bread' (Schmookler 1966, pp.208).

At first sight, this point of view appears to be incompatible with the Schumpeter I model discussed before, which strongly stressed the role of technology-push effects. In the next sections however, it will be argued that the relevance of both concepts may again be related to different phases of the life cycles of new industries and technologies.

Another important issue nowadays appears to be the question whether *innovations are clustered in time*. Can some periods be characterized as 'spring-tides' of innovative activities? In this respect Schumpeter (1939) already stressed the impact of 'bunches' of innovations to long wave economic fluctuations, i.e., the Kondratiev cycle. Kuznets (1940) however, in reviewing Schumpeter, seriously questioned the fact whether innovations would really disturb the economic growth pattern in a significant way. This would only be the case when *basic innovations* - i.e., innovations which influence the existing economic system in a tremendous and prolonged way - emerge *discontinuous* in time and/or when *further improvement innovations*- i.e., the innovations generated during the *diffusion or 'swarming' processes* following the introduction of basic innovations (see below)- would be *situated together in time*.

It has been precisely along these lines that subsequent research efforts, aiming at the explanation of long wave economic fluctuations by means of 'spring-tides' of innovative activities, have prolonged. The more current popularity of the long wave Kondratiev cycle can partly be ascribed to the fact that several research efforts indeed 'support' this *clustering* of basic and/or improvement *innovations in time*.

On the one hand, some researchers have especially concentrated upon the *first* issue raised by Kuznets (see, for example, Mensch 1981a, 1981b, Kleinknecht 1981, 1987a). Although not denying the relevance of the second issue raised by Kuznets, their (implicit) supposition is that *both processes* are *sequentially interrelated* (cf., Mensch et al 1981). So in their interpretation, the implicit hypothesis is that the 'swarming' process of further improvement innovations will soon follow the initial introduction of the relevant basic innovations.

On the other hand, Clark et al (1981a, 1981b) and Freeman et al (1982), for example, especially stress the second issue raised by Kuznets:

> We stress the importance of the 'diffusion' process and the way in which a series of further innovations are generated as a swarms of imitators move in to invest in a new technology, attracted by the exceptionally high profits (Freeman et al 1982, pp.xl).

In their interpretation, the diffusion process 'may not necessarily occur immediately after a basic innovation although it may do so if other conditions are favourable' (Freeman et al 1982, pp.65). These other 'conditions' also refer to non-economic - e.g., political and socio-institutional forces. They will be discussed in later sections in the context of new 'technology systems'. This also applies to the issue of the 'bunching' of (basic) innovations in time.

First of all however, in the next section we will briefly discuss various types, results and methodological problems inherent to those *empirical* studies, that try to provide evidence for the relationship between innovative activities and economic performance. This will also set the stage for the empirical analysis in the second part of this study.

1.3 MEASURING INNOVATIVE ACTIVITIES AND ECONOMIC PERFORMANCE

An important problem often encountered in the empirical testing of the relationship between innovative activities and economic performance lies in the *measurement* of both concepts. Concerning the measurement of the first concept, i.e., innovation, one may make a distinction between efforts to measure the *input* aspect - labelled innovation potential here - and the *output* aspect - labelled innovativeness here - of the innovation process. As to the *input* dimension, indicators such as number of R&D employees, R&D expenditures, number of engineers employed and so on, are often used. For the *output* dimension indicators such as number of patents granted, number of new products generated, number of new process technologies implemented and so on, are often used.

Especially the measurement of the output aspect appears to be an arduous task because of lack of relevant and reliable data (cf., Kleinknecht 1987b, and Hansen 1986). For this reason, most empirical studies approximate the level of innovative activities in a firm, sector or country by means of (a mixture of) *input* indicators. Next, these indicators are linked to several measures of relative economic performance. In the sequel we will present some examples and problems concerning the latter type of analysis.

Historically, the relation between technological change and economic performance has often been studied by means of the determination of the correlation between indicators of R&D expenditures and the growth of (total) *factor productivity* (cf., Kendrick 1961, Mansfield 1968). These studies especially concentrated on the manufacturing sector. In general, the findings pointed at a *positive correlation* between technological change (approximated by R&D indicators) and economic performance (i.e., growth in total factor productivity).

However, as Nelson and Winter (1977) argue, the *direction of causality* may often appear ambiguous. Consequently, one of the main problems inherent in this type of analysis is a *specification* problem:

> Obviously, there are a tangle of causations, from R&D to productivity growth, from productivity growth and lower prices to growth of output, from growth of output in the presence of scale economies to productivity growth, from expansion of the industry to greater incentives to R&D and so on (Nelson and Winter 1977, pp.45).

Furthermore, it is clear that not all R&D activities are aimed at, or reflected in, increased factor productivity. On the other hand, growth of total factor productivity may also result from *external R&D activities* embodied in the capital inputs of a firm or sector. In this respect, Nelson and Winter (1977), in discussing Terleckyj's (1974) findings that in a multiple regression equation own R&D appeared to be rather insignificant in 'explaining' total growth of factor productivity compared to external R&D embodied in capital inputs, point at one possible explanation for this finding:

> One possibility is that research and development spending financed by an industry is largely focused on new product design that, because of weaknesses in price indices, does not show up adequately in measures of growth of industry output; in contrast process improvement, which show up more reliably in increased productivity, stem largely from improved input and capital equipment (Nelson and Winter 1977, pp.44).

Approximating the relationship between technological change and economic performance, in the way sketched above, may be further hampered by the fact that the *productivity of R&D efforts may differ across sectors and time*. In this respect, Ayres (1987), for example, remarks:

> technological progress is not simply proportional to aggregate R&D invested. In particular, the probability of a major innovation depends very critically on the state of the supporting technology at the time (Ayres 1987, pp.15).

As will be discussed in the next sections, several other authors point at the tendency of the productivity of R&D efforts to decline during the 'life cycle' of new technologies.

So, although these empirical findings in general suggest a positive correlation between R&D intensity and growth of total factor productivity, the exact interpretation of these results still remains hazardous. This applies especially to the issues of *causality* as well as to the *changing nature of the relations* across sectors and time.

More recent research efforts have tried to link R&D intensity to other 'performance' indicators, like percentage of sales attributable to new products, export orientation, employment growth, and growth of sales. Much of the criticism raised above applies, mutatis mutandis, also to these types of studies. As an example we will briefly discuss a table with empirical results derived from Hansen (1986). On the basis of an inquiry among 600 large manufacturing firms in the US the following table can be constructed from Hansen's analysis.

Table 1.1 R&D Intensity and Sales of New Products

SIC	Typology	R&D/Net Sales (%)	% of firms having more than 20% of net sales from new products	% of firms hav less than 10% net sales from new products
38		5.7	77	3
35	technology	5.0	52	16
37	intensive	4.1	79	20
36		3.4	66	21
28		3.1	9	79
34	manufacturing	2.5	78	20
30	industries	2.3	70	27
32		1.7	0	100
26		1.3	0	47
33	raw material	1.3	20	79
13	heavy industry	0.5	55	43
20	light	0.8	10	56
22	manufacturing	0.3	38	62
	all firms	2.5	38	49

Derived from: Hansen (1986)

In this table economic performance has been approximated by means of the share of new products - i.e., products first marketed in the period 1978-1982 - in total net sales of manufacturing firms. In general, the results in column 4 and 5 display the reverse pattern. However, both results point at the fact that more technology intensive firms (as measured by the R&D-sales ratio) are more successful in terms of a 'revitalisation' of their market outlets.

A final example concerning this type of empirical research is formed by the currently quite popular 'high tech' studies (see for example, OTA 1984, Bouwman et al 1985a, 1985b, Markusen et al 1986). In these studies, the 'high tech' sectors are usually selected on the basis of various criteria referring to the *input* aspect of innovative activities. Next, the economic performance of the selected sectors is compared with other (non-'high tech') sectors.

An example of the latter type of study can be found in Markusen et al (1986) who define 'high tech' sectors on the basis of the occupational structure. They approximate economic performance by means of the relative growth of employment (i.e., in comparison to the non-high tech sectors). Although not all selected high tech sectors exhibited growing employment figures (pointing at sectoral variation), in the period 1972-1981 they conclude: 'The 30 top-growing high tech industries ... accounted for 99% of all net high tech employment growth and 87% of all manufacturing job growth' (Markusen et al 1986, pp.31).

So the general pattern of the above-mentioned empirical studies supports the view that innovative activities and economic performance are *positively* correlated. Those firms, sectors or countries that remain at the technological frontier appear to perform better than the technological laggards.[1]

The acknowledgement that innovative activities do not take place in an isolated vacuum, but shape and transform industries as well as the other way around, and that these impacts may change during the life cycles of new technologies and industries has been the subject of many recent research efforts. In this context, Freeman et al (1982), for example,

> argue for the importance of technical change both as a stimulus to an upswing in investment activity - an 'engine of growth', and as a response to the changing pattern of demand and prices - a 'thermostat' (Freeman et al 1982, pp.17).

We will discuss this issue in the next sections. In this context, we will take a *long term* point of view and sketch some of the ideal-typical (economic and technological) regularities concerning the emergence and maturity process of *industries and technologies*.

1.4 NEW 'TECHNOLOGY SYSTEMS' AND TECHNOLOGICAL 'TRAJECTORIES'

In the foregoing sections, we briefly pointed at a number of basic issues concerning the (determination of the) relationship between innovative activities and economic performance. In this context, certain aspects of this relationship still remained unclear. More recent research efforts however, have clarified part of this cloudiness by considering the interdepencies between *life cycles of industries and technologies*. These efforts will be considered and integrated in the following sections.

One of the basic propositions dealt with by Dosi (1984) when studying technological change is that:

> The evolution of technologies through time presents some significant regularities and one is often able to define 'paths' of change in terms of some technological and economic characteristics of products and processes (Dosi 1984, pp.12).

The identification of such, albeit ideal-typical 'paths' of technological and economic change, underlies precisely the purpose of the remaining part of this chapter. For this purpose, we will start with a discussion of rather abstract, but useful concepts like new *'technology systems'*, technological *'paradigms'* and technological *'trajectories'*.

The first concept deals with the *discontinuous* emergence of new life cycles of industries and technologies. In this respect Freeman et al (1982) define new 'technology systems' as:

> the 'clusters' of innovations ... associated with a technological web, with the growth of new industries and services involving distinct new groupings of firms with their own 'subculture' and distinct technology, and with new patterns of consumer behaviour' (Freeman et al 1982, pp.68).

In their interpretation new *upswings* in economic development crucially depend on the 'emergence' of such new 'technology systems'. As regards the 'explanation' of the emergence of such a new 'system', the causal factors exceed the economic dimension, however. This emergence will depend on the complex interplay between *scientific, socio-*

institutional and economic mechanisms or innovations (see also below).
New 'technology systems' emergence *infrequently* in time as these
constituent forces - and the interplay between these forces - are
discontinuous by nature.

As regards the *scientific* mechanisms, Ayres (1987), for example,
has pointed at the irregular impact of advances in science and
technology upon innovative activities. This impact is discontinuous
because it is characterized by *barriers* and *breakthroughs* as reflected
in the 'bunching' of basic innovations in time (cf., Mensch 1981a, van
Duijn 1983 and Kleinknecht 1987a, 1987c). In this respect, some well-
known historical examples of basic innovations are: the steam engine,
iron-making, internal combustion engines, aircraft (cf., Ayres 1987),
nylon, radio, television, integrated circuits (cf., Freeman et al 1982).

Perez (1983), among others, has pointed at the relevance of the
socio-institutional framework to the techno-economic framework. In her
interpretation economic depressions are generally characterized by a
'mismatch' between both frameworks. New technology may only 'take-off'
if the socio-institutional framework is receptive to it. Consequently,
also socio-institutional changes (i.e., innovations) are essential for
the economic effects - to be derived from such radical changes in
science and technology (i.e., basic innovations) - to be effectuated.[2]

Also the *economic incentives* to invest in new technology may be
fluctuating over time. In this respect, for example, van Duijn (1981)
and Mensch et al (1981) argue that these incentives may be related to
different phases of the long wave Kondratiev cycle. Mensch et al (1981)
postulate

> an increase in the propensity of capital owners for investing their
> accumulated capital in alternative technologies at times of
> discontinuity of prosperity of old lines of business and commerce
> (Mensch et al 1981, pp.173).

Several other long wave theorists have also tried to endogenize
the economic long wave by pointing at such differential economic
incentives to innovate during different phases of the long wave (see,
for instance, Kleinknecht 1981, 1987a, Vasko 1987). It would be beyond
the purposes of this chapter however, to discuss these various studies

in detail.

In case a new 'technology system' becomes established, the wider 'diffusion' of the many *new types of products and services* - to be *derived* from the underlying constellation of (prior) *basic and socio-institutional innovations* - will really 'take off'. This will result in a 'booming phase' of new life cycles of industries and technologies.[3] This issue will be further discussed in the next section.

As regards the further innovative activities related to these new types of products and services it is useful to apply the notion of technological 'trajectories' (cf., Dosi 1984) or 'natural trajectories' (cf., Nelson and Winter, 1982). Essentially, this concept refers to the fact that technological change - concerning new types of products and services - is *not* a purely *random and static* diffusion process, but further proceeds along specific 'paths'. Consequently, we will define new *technological 'trajectories'* as the 'paths' of further innovative activities underlying new types of products and services (see also subsection 1.5.3).

In this respect, the purposes and direction of these innovative activities along such 'paths' are often well-defined and proceed along underlying principles and guideposts (cf., Sahal 1981; see also the discussion of technological 'paradigms' below). In section 1.5.3 we will discuss some ideal-typical regularities as regards the *types* of innovative activities performed along such 'trajectories'.

As an example of such 'trajectories', Dosi mentions the semi-conductor sector, in which technical progress along the established 'trajectories' especially aims at 'minutarisation (increasing density of the circuits), speed, reliability and decreasing costs' (Dosi 1984, pp. 67). Nelson and Winter (1977) present the following examples of 'natural trajectories':

> In jet engine technology, thermodynamic understanding relates the
> performance of the engine to such variables as temperature and
> pressure at combustion. This naturally leads designers to look for
> engine designs that will enable higher inlet temperatures, and
> higher pressures. In airframe design, theoretical understanding (at
> a relatively mundane level) always has indicated that there are
> advantages of getting a plane to fly higher where air resistance is
> lower. This leads designers to think of pressurizing the cabin,
> demanding aircraft engines that will operate effectively at higher
> altitudes (Nelson and Winter 1977, pp.58).

As pointed out by Nelson and Winter, 'natural trajectories' may be *specific* to a particular technology although some apply to a *wide range* of technologies. Examples of the latter trajectories are 'progressive exploitation of latent scale economies and increasing mechanization of operations that have been done by hand' (Nelson and Winter 1977, pp.58). In the twentieth century two important new 'natural trajectories' emerged, i.e., 'the exploitation of understanding of electricity and the resulting creation and improvement of electrical and later electronic components, and similar developments regarding chemical technologies' (Nelson and Winter 1977, pp.58).

Some *general guidelines* - applying to several 'trajectories' at the same time - concerning the direction of 'normal' technological progress along such 'trajectories' will be provided by the prevailing *technological 'paradigm'*. In a similar vein like Kuhn's scientific paradigms, Dosi defines a 'technological paradigm' as: 'a 'model' and a 'pattern' of solution of selected technological problems, based on selected principles derived from natural sciences and on selected material technologies' (Dosi 1984, pp.12).

Nelson and Winter (1977) define a similar 'umbrella' concept of *'technological regimes'*, which defines several 'natural trajectories'. They describe this concept as follows:

> The sense of potential, of constraints, and of not yet exploited opportunities, implicit in a regime focuses the attention of engineers on certain directions in which progress is possible, and provides strong guidance as to the tactics likely to be fruitful for probing in that direction. In other words, a regime not only defines boundaries, but also trajectories to those boundaries (Nelson and Winter 1977, pp.57).

The similarity with Dosi's concept of technological 'paradigm' will be clear from this definition.

In concluding this section and in view of the purposes of this study, the following basic concepts and ideas will be adopted from the recent innovation literature:

1. New 'technology systems' emerge *infrequently in time* as a result of the complex interplay between scientific, socio-institutional and economic forces (innovations).

2. In case a new 'technology system' becomes established this will result in a *'bunching'* of the 'take-off' of new technological 'trajectories' to be derived from the underlying constellation of basic (and socio-institutional) innovations.

3. This opening-up of many new 'trajectories' will result in 'swarming processes' of (new) 'Schumpeterian' firms attracted to - and trying to move further on - these 'trajectories'. These 'swarming' processes will result in new *life cycles of industries and technologies*.

A more thorough analysis of the first issue would be beyond the purposes of this study. It will be clear from the foregoing discussion that the 'explanation' of the emergence of a new 'technology system' is a rather complex issue encompassing, inter alia, socio-institutional and economic dimensions.

Although researchers may disagree concerning the exact 'explanation' and historical timing of such new 'technology systems', concerning the *second issue* it is nowadays quite commonly agreed upon that the economic 'boom' phase following World War II, largely depended upon the fact that the take-off of several new 'trajectories' was rather concentrated in time. Underlying these 'trajectories' were basic innovations such as the computer, drugs, transistor and plastics.

In more recent decades however, the shortage of new 'trajectories', combined with a decline in the possibilities to generate further product innovations along previously established 'trajectories', had a negative impact upon manufacturing employment (cf., Van Duijn 1981, Freeman et al 1982, Rothwell and Zegveld 1985).

Promising *new 'trajectories'* of today appear to based upon basic innovations in the field of composite materials, artificial intelligence, biotechnology (cf., Ayres 1987), information and telecommunication technology (cf., Naisbitt 1984, Freeman 1987a, 1987b, Orishimo et al 1988).

In the following sections we will concentrate on the *third issue* mentioned above, i.e., a description of some of the ideal-typical *economic* and *technological regularities* related to the 'swarming processes' mentioned above.

1.5 ECONOMIC AND TECHNOLOGICAL ASPECTS OF THE 'SWARMING' PROCESSES

1.5.1 Life Cycles of Industries and Technologies

In the foregoing section, we discussed the *discontinuous*, and partly exogenous (i.e., to the economic domain) nature of the emergence of new 'technology systems' and the related 'bunching' of the 'take-off' of new technological 'trajectories'. We also pointed at the fact that the *economic* effects of basic (and socio-institutional) innovations-which underlie these new 'trajectories' - will be effectuated through 'swarming' processes of (new) 'Schumpeterian' firms attracted to these 'trajectories'. In this respect, Freeman et al (1982) remark:

> The swarms which matter in terms of their expansionary effects are the diffusion swarms after the basic innovations and the swarming effects associated with a set of interrelated basic innovations, some social and some technical, and concentrated very unevenly in specific sectors (Freeman et al 1982, pp.67).

These 'swarming' processes will result in *new life cycles of industries and technologies.*

As to the *spatial* dimension especially the larger metropolitan complexes are often considered to provide the incubation conditions necessary for the take off of such new life cycles. The spatial dimension of such new economic activities will be dealt with more thoroughly in subsequent chapters of the present study. In this respect in Chapter 9, for example, the incubation potential of metropolitan areas with respect to relatively new and innovative (sub)sectors-linked to the more recent information 'technology system' - will be considered for the Dutch context.

First however, in the following sections we will concentrate upon a further discussion of some (ideal-typical) economic and technological[4] regularities of the 'swarming' processes mentioned above. In this context, our analysis will be biased towards more recent research efforts. We will start with the economic dimension.

1.5.2 Economic 'Regularities' of the 'Swarming' Processes

Especially in initial phases new technological 'trajectories' will often provide favourable profit opportunities. Consequently, many (new) 'Schumpeterian' firms will be attracted by the potential high profits related to these 'new avenues'. At this time *economic growth* - in terms of employment, sales, investments and so on - will be *booming* in these new industries.

Firstly, concerning the *supply* side, this high profit potential may be due to the fact that initially the *'productivity' of R&D efforts* - i.e., in terms of the generation of (further) innovations - will be the *highest* just after the introduction and social 'receptiveness' of the underlying basic (and socio-institutional) innovations (see, for example, Wolff 1921, Mensch 1981b, Metcalfe 1981, Ayres 1987, Kleinknecht 1987c). In this respect, Metcalfe (1981) states that:

> the scope for improvement in any technology is bounded, and approach to this limit is slowed down by the increasing difficulty in generating successive incremental improvements. As a corollary to this, the more significant improvements tend to cluster close to the original time of innovations' and 'the cumulative impact of the secondary will often exceed the impact of the initial innovations. Retardation in the rate of secondary innovation is thus a dominant factor in the overall tendency to retardation in the growth of a new industry (Metcalfe 1981, pp.351-352).

Also Ayres (1987), in discussing empirical studies concerning the productivity of R&D efforts, concludes that there exists 'strong evidence of the tendency of R&D productivity to decline in the mature phase of the life cycle of a product of process' (Ayres, 1987 pp.34-35).

This decreasing productivity of R&D also underlies the before-mentioned shift from an energy-intensive mass-production paradigm to an information-intensive paradigm. The first paradigm 'was reaching limits of productivity and profitability gains, due to a combination of exhaustion of economies of scale, erosion of profit margins through 'swarming', market saturation in some sectors, diminishing returns to technical activities (Wolff's law) and cost pressures on input prices'. The new paradigm, however 'offers the possibility of renewal of productivity gains and increased profitability' (Freeman 1987b, pp.305)

Secondly, concerning the *demand* side, the initial generation and further improvements of the new types of products and services - along the newly established 'trajectories' - will in general exhibit the *largest increases in marginal utilities* (cf., Mensch 1981b, Sahal 1981). Subsequent improvements will often be governed by 'some law of diminishing marginal utility' (Mensch 1981b, pp.172). In the initial phases preferences of the buyers will be largely *unknown and volatile*. Entrepreneurs attracted to the new 'trajectories' will try to capture part of the various growing markets by generating - or further improving - new types of products and services (i.e., by generating further product innovations; see subsection 1.5.3).

As a consequence of these demand and supply conditions prevailing in initial phases the *structure of supply* is 'often characterized by a high rate of birth and mortality of new 'Schumpeterian' companies enjoying oligopolies on clusters of innovations' (Dosi 1984, pp.93). In these phases the *Schumpeter I model*, as discussed in section 1.2, will be highly relevant.

The *empirical validity* of the Schumpeter I model during the initial phases of industry life cycles has, for example, been demonstrated with respect to the semi-conductor, chemical and electronics sector (cf., Freeman et al 1982, Dosi 1984). Freeman et al (1982), for example, demonstrate that in the 'recessive' period 1975-'80 the very fast growing firms in the US (in terms of sales) appeared to be relatively small firms 'within the broad technological area of information processing and electronics, which has often been heralded as the new technological cluster around which a major new upswing might be set in motion' (Freeman et al 1982, pp.143). This provides evidence concerning the proposition that 'long-term cyclical upswings might be more closely associated with a resurgence of mark I small firm innovators' (Freeman et al 1982, pp.143). We will return to this issue in the next section.

As the 'bandwagons' along the new 'trajectories' proceed however, and more entrepreneurs enter the scene, several of the new 'trajectories' will be confronted with *market saturation*. Consequently the rate of 'Schumpeterian' new firm formation will level off, and *economic growth* in these new industries will become *stagnant*. This may especially apply to the employment variable (resulting from labour saving capital 'vintages').

Also in these later phases the profitability of generating further product innovations will level off as the concomitant costs increase coupled with a decreasing marginal utility of further improvements (see above). So, in these phases products will become more standardized (e.g., all producers of videos applying the same VHS system) and *price competition* will become a more effective means of competition. Consequently, as 'normal' technological progress along newly established 'trajectories' proceeds, the emphasis of the innovative activities will increasingly shift toward improving *production efficiency* (see also below).

In this respect, price competition between firms may result in more *oligopolistic market structures*. The extent to which such a market structure will emerge depends, inter alia, upon the possibility to realise economies of scale in large scale production units as well as the relative success of firms to reduce production costs by means of process innovations (cf., Kamien and Schwartz, 1982). These forces may increase the relative importance of the Schumpeter II model in these later phases.

Another 'regularity' as regards the innovative activities performed along (new) technological 'trajectories' appears to be the shifting emphasis from *'technology push'* to *'market pull'* generated innovations. Initially, the new technologies are still immature and experimental and considerable efforts will be devoted to control the technology (and its applications) itself. Consequently, especially in the initial phases of the 'take-off' of new 'trajectories' the *'technology-push'* argument will be highly relevant. So, in more recent periods the latter argument appears to be of utmost importance in generating new 'trajectories' in the field of micro-electronics, bio-technology, composite materials and telecommunication technology (see also Dosi 1984, Freeman 1987a 1987b, Ayres 1987, Roobeek 1988).

As technological progress along the newly established 'trajectories' proceeds, however, the *market* will become increasingly dominant in determining the further pace and direction of innovative activities. So van Duijn (1981) concludes: 'Schmookler's empirical findings have to be placed in their proper perspective: within established sectors, technological change will follow increases in market demand' (van Duijn 1981, pp.274). Ayres (1987) summarizes this shifting emphasis as follows:

> In the life-cycle of a technology, the balance between push and pull changes over time. In the very early period, technology push

can sometimes be quite important... Later in the life cycle pull
takes over. Its function is to induce a collection of competing
entrepreneurs to find an optimum balance between product
performance and price for the customer vis-a-vis profitability for
the producer... In summary, the importance of push is likely to be
highest at the very beginning of the life cycle (Ayres 1987, pp.43-
44).

The same arguments can also be found in Dosi (1984) and Kamien and
Schwartz (1982). Kamien and Schwartz consider the technology-push and
demand-pull hypotheses to be complementary 'with the former being more
of a long-run theory and the latter, a short-run theory' (Kamien and
Schwartz 1982, pp.36).

After this brief discussion of some economic aspects of the
'swarming' processes along newly established technological
'trajectories', we will turn to the technological dimension in the next
subsection.

1.5.3 Technological 'Regularities' of the 'Swarming' Processes

As discussed in the previous subsection, the 'swarming' processes
of (new) 'Schumpeterian' firms attracted to new 'trajectories' will
result *both* in generating and further improving of new products and
services and related process technologies. In this respect, the
'swarming' processes along new 'trajectories' can be considered as a
kind of '*learning process*' - i.e., a 'creative destruction' process - in
which (new) 'Schumpeterian' firms will constantly try to move further on
these 'trajectories'. In this context, Clark et al (1981, pp.153), for
example, remark:

> diffusion is seldom the simple imitation of a homogenous product of
> process by a succession of enthusiastic firms rushing in to exploit
> a technical innovation. More commonly major improvements and even
> further related basic innovations are made when the band-wagon
> starts to roll (Clark et al 1981b, pp.153).

The same arguments can, for example, also be found in Kuznets's
earlier work (1930), and more recently in Rosenberg (1976), Metcalfe
(1981), Freeman et al (1982) and Granstrand (1986). In this study we
will label these further improvements the '*creative diffusion*' process,

as opposed to those diffusion models and theories which consider the diffusion process to be a static 'carbon copy' process of imitation (cf., Rosenberg 1976).[5]

In this respect, Nelson and Winter (1977), point out that Mansfield's empirical study (Mansfield 1968) has presented considerable support to such a 'creative diffusion' process:

> perhaps his most interesting results involve comparisons of firm growth rates, where he finds that innovating firms in fact tend to grow more rapidly than the laggers. However, while the advantage of the innovator tends to persist for several periods, the advantage tends to damp out with time, apparently because other firms have been able to imitate, or to come up with comparable or superior innovations (Nelson and Winter 1977, pp.65).

Sahal (1981), in promoting a 'system's view of technology' - which stresses *the changes in product characteristics* inherent in these further innovative activities - makes a similar point:

> In the literature on the subject it is generally assumed that the characteristics of an innovation do not change during the course of its adoption. In reality, however, changes in product characteristics often lead to new markets and expansion of the old, thereby significantly altering the scope for adoption of innovation (Sahal 1981, pp.16).

Metcalfe (1981), in criticizing the static nature of the standard epidemic diffusion models, states that in these models 'a given innovation is diffused within an unchanging adoption environment, although there are well documented reasons for expecting both innovations and environment to change as diffusion proceeds' (Metcalfe 1981, pp.349).

An important corollary of this conceptualisation lies in the fact that extraordinary profits to be derived from innovations (whether product or process) may only be *temporary*. Other entrepreneurs, attracted by the exceptional profit possibilities in the initial phases of the 'swarming' processes, will try to gain a competitive edge by means of further moving along the underlying technological 'trajectories'.

As (implicitly) stated in the discussion of some economic aspects of the 'swarming' process in the previous subsection, in the course of time the *main emphasis* of further innovative activities along newly established 'trajectories' will shift from *product to process oriented innovations*. In the sequel, this will be labelled the '*innovation life cycle*' concept.

This concept has been described, for example, by Abernathy and Utterback (1978), Sahal (1980, 1981) and Rothwell and Zegveld (1985). According to this concept, in the initial phases of the 'swarming process', the underlying basic (and socio-institutional) innovations will especially result in *generating* and *further improving* new products and services. In this phase, the new technology is still unstable and experimental, and the characteristics of the market are still largely unknown. Producer-user interrelations will result in several modifications of these products and services (cf., for example, the various video-systems that initially existed). Consequently, in the initial phases *competition* will largely be based upon *product innovation*.

As learning along new technological 'trajectories' proceeds however, and demand patterns become more crystallized and known, the new products and services will become more '*standardized*', i.e., become closer substitutes. As regards several 'trajectories' some 'dominant design' will emerge (cf., Sahal 1981). In this respect, it will become increasingly difficult and costly to generate further product innovations. In this context, Kamien and Schwartz (1982) stress the fact that 'the incentive to develop a cost-reducing innovation is greater for a product with close substitutes than for one with few substitutes' (Kamien and Schwartz 1982, pp.43).

Consequently, as 'normal' technological progress along new 'trajectories' proceeds, the *emphasis* of the (further) innovative activities will increasingly shift towards the generation of improved (cost-efficient) production techniques, i.e., *process innovations*. These innovations aim at increasing the price-competiveness of (new) firms concerning products (or services) that have become more or less 'standard' by that time.

The 'innovation life cycle' concept is depicted below in Figure 1.3.

24

Figure 1.3 The Innovation Life Cycle Concept

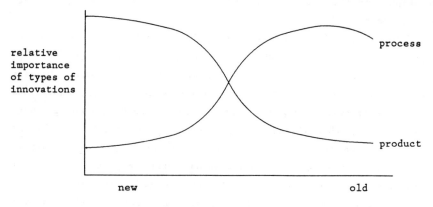

relative
importance
of types of
innovations

process

product

new   old

age of technological 'trajectory'

This shift in emphasis from product to process technologies is also implied by the long wave theory provided by Mensch et al (1981) and van Duijn (1981), for example. One of the four hypotheses Mensch et al (1981) postulate concerning 'an innovation-oriented long-term theory of industrial evolution' is that 'Over time, a substitution of process innovations for product innovations will take place' (Mensch et al 1981, pp.284). The same arguments can also be derived from van Duijn's 'model of the propensity to innovate' during different phases of the long wave. In presenting some innovation characteristics of the more recent recessive period of the seventies, van Duijn (1981, pp.270) states that 'there is indeed a lack of (employment-creating) product innovations in new sectors, but certainly no lack of (labour-saving) process innovations in existing sectors'.

Rothwell and Zegveld (1985), in their 'Model of Postwar Industrial Evolution', point out the relevance of the 'innovation life cycle' at the level of the manufacturing sector as a whole. During the period 1945-1964, which they label 'the dynamic growth phase' there was 'emphasis on product change and the introduction of many new products' (pp.33), while 'competitive emphasis is mainly on product availability and non-price factors' (pp.33). The period of the late sixties to date however, which they label the 'maturity and market saturation phase' is characterized by 'some product change, but emphasis is predominantly on production process rationalization' (pp.33) and 'where products are

little differentiated the importance of price competition is high'
(pp.33).

The empirical validity of this concept has, for example, been
confirmed by Freeman et al (1982) who classified 195 radical innovations
(relating to 60% of the manufacturing sector in the UK) according to
date and *type* of innovations. Figure 1.4 below depicts the observed
pattern:

Figure 1.4   Number and Types of Radical Innovations in British Industry

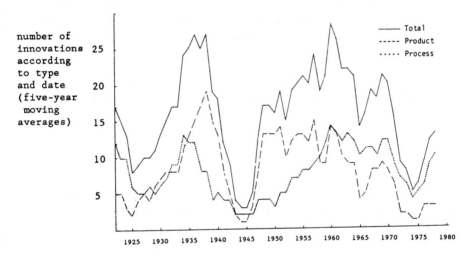

Source: Freeman et al (1982), pp.52

Especially concerning the period 1945-1980 the 'innovation life
cycle' concept, as discussed above, appears to be relevant. These
findings also comply with the more qualitative statements of van Duijn
(1981) and Rothwell and Zegveld (1985) concerning the *general
characteristics* of post-war innovative activities in the manufacturing
sector.

At the sectoral level, Freeman et al (1982) discuss the
(international) growth cycle of the synthetic materials industry (see
also Freeman et al 1968). Concerning the innovative activities performed
within this sector, they conclude:

Although we have stressed that basic innovations continued to occur in the 1950s and 1960s, and that the rate of secondary and induced invention continued at a very high rate, there were indications of diminishing returns to further investment in R and D and other technical activities and of a marked shift to process innovations (Freeman et al 1982, pp.99).

Coombs and Kleinknecht (1984) analyzed a dataset on 500 innovations concerning the period 1953-1973, constructed by Gellman (1976). Their analysis confirmed that product innovations were heavily concentrated within the fast growing industries (affirming that initially innovative activities in new growth sectors will be biased towards product innovations). During the period considered however, a shift occured within these growth industries *from product to process innovations*. Kleinknecht (1987b) summarizes these findings as follows:

The shift from product to process innovation within highly innovative industries during the postwar upswing fits the idea of an industry life cycle, suggesting that, with an increasing degree of maturity, growth industries lose their ability to generate employment (Kleinknecht 1987c, pp.231).

The same conclusion can also be derived from the Mahdavi (1972) list of 'important' innovations (cf., Kleinknecht 1981).

Dosi (1984) has made a detailed study of the semi-conductor sector. His findings also confirm the *shifting emphasis from product to process* (capital-equipment) technology. 'In the overview of structural evolution of the industry, one observed the emergence, especially in the first two decades of history of the industry, of 'Schumpeterian' companies often associated with the introduction of new products' (Dosi 1984, pp.193). As the 'swarming' processes proceeded, however, 'The process of establishing a certain technological path appears to imply, broadly speaking, a trend towards the increasing incorporation of technology into capital equipment' (Dosi 1984, pp.193).

Consequently, the innovation life cycle concept can be considered as one of the ideal-typical 'regularities' of innovative activities along newly established 'trajectories'. In subsequent chapters we will also try to judge the empirical validity of this concept (see in particular Chapter 8 of this study). First however, in a concluding section the most important findings of this chapter will be summarized.

1.6 RETROSPECT

In the foregoing sections we have concentrated on the 'linkages' between life cycles of industries and technologies. In this context we discussed briefly various concepts related to these 'linkages'.

First of all, concerning the issue of the *impact* and *timing* of the *economic effects* of basic and socio-institutional innovations, we discussed in section 1.4 the concept of new *'technology systems'* (cf., Freeman et al 1982). In this respect, we discussed that the complex (and discontinuous) interplay between scientific, socio-institutional and economic factors may lead to 'a constellation of circumstances favourable to the exceptionally rapid growth of one or more industries, each involving the combination of a number of related inventions, innovations and social changes' (Freeman et al 1982, pp.70).

The 'emergence' of a new 'technology system' will lead to (a 'bunching'of) the 'take-off' of new technological 'trajectories' related to the constellation of basic - and socio-institutional - innovations. The 'swarming' processes of (new) 'Schumpeterian' firms attracted to these 'trajectories' - generating and further improving new products and services as well as related process technologies - will result in new life cycles of industries and technologies. Figure 1.5 below summarizes the framework sketched above.

Figure 1.5 New Life Cycles of Industries and Technologies

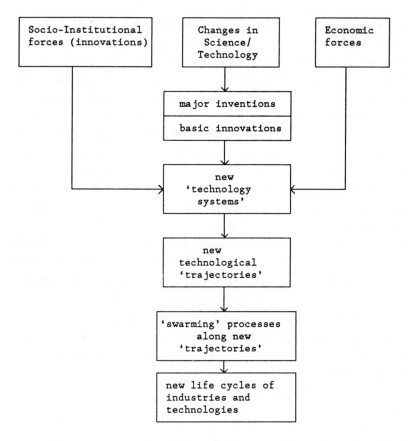

For our purposes, the most important (economic and technological) 'stylized facts' inherent in these 'swarming' processes are the following.

Life cycles of industries and technologies are interrelated. In this respect, Freeman et al (1982) remark: 'The important phenomenon to elucidate if we are to make progress in understanding the linkages between innovations and long waves is the birth, growth, maturity and decline of industries and technologies' (Freeman et al 1982, pp.65). The 'swarming' processes - i.e., the entrance of many 'imitators' and the accompanying further innovative activities - are fundamentally the development of both industries and technologies. In this respect, Dosi remarks: 'the expansion of the innovative firms and imitations by other firms, is intimately associated with additional innovations and improvements' (Dosi 1984, pp.287).

Consequently, the 'swarming' processes are *'creative'* processes and not simple 'carbon-copy' processes of imitation. New 'imitator' firms will try to become competitive by further moving along the underlying technological 'trajectories'. As the 'swarming' processes proceed, *new products* will be generated and further improved *continuously*, although at a *decreasing rate*. These further innovations are often a necessity to enter the market : 'asymmetric innovative capabilities associated with some limit-price force potential entrants to introduce further innovations in order to enter the market. This same process tends to improve the performance characteristics of the innovative commodities and to widen the universe of potential adopters' (Dosi 1984, pp.286). Especially in the initial phases of the 'swarming' process, competition by means of *product innovations* will be an effective means of competition as both the new technology as well as consumer preferences are still immature and experimental.

Concerning the *Schumpeter I and II model of innovation*, it can be stated that in the *initial phases* of the 'swarming' processes the *Schumpeter I model* will often be highly relevant. In these phases, many (new) firms will be attracted to the scene because of the high profit potential of the new 'trajectories'. In such periods entry-barriers will be low as no one controls the entire market yet. In these phases empirical research has indicated that economic opportunities are very favourable to entrance, as both the *costs* of further innovations will be lowest and the *expected returns* of R&D highest, witness also recent developments with respect to CAD/CAM systems (cf., Mensch 1981b, Freeman et al 1982, Dosi 1984, Kleinknecht 1987c, Ayres 1987 and Roobeek 1988). In this respect, the Schumpeter I model appeared to be highly relevant during the initial phases of the electronics, chemical and the semi-conductor industry (cf., Freeman et al 1982, Dosi 1984).

As more entrepreneurs enter the scene however, and consequently market saturation sets in, the rate of Schumpeter I new firm formation will level off. As discussed above, both costs and expected returns on further improvements will diminish the incentives to entrance. 'In economic language, the marginal product of R&D in any established field tends to decrease' (Ayres 1987, pp.28). This pattern can, for example, be observed in the automobile and aircraft industry where it becomes increasingly difficult to increase the energy efficiency. Accordingly, *price competition* will become a more important means of competition.

The number of firms operating in this sector may diminish and a more oligopolistic market structure may be generated depending inter alia upon the extent to which the 'classical trajectories' (cf., Nelson and Winter 1977) of obtaining economies of scale are relevant to the specific sectors and technologies considered.

In the course of the 'swarming' processes along the newly established 'trajectories', emphasis will shift from *technology induced* innovations to *demand pull* innovations. In the early phases of the 'take-off' of the new 'trajectories' considerable efforts will be devoted 'to make it work', i.e., to master the new experimental technology itself. As demand patterns become more known and established however, the *market* will become increasingly important in determining further innovative activities. 'Exogenous science and new technology tend to dominate in the early stages, whilst demand tends to take over as the industry becomes established' (Freeman et al 1982, pp.38). Consequently, the relative importance of the demand-pull versus technology push hypothesis will shift in the course of the 'swarming' process.

A similar statement can be made concerning the relevance of *product* versus *process* innovations. As technological progress - or the 'swarming' processes - along the new 'trajectories' proceeds, the main emphasis of the further innovative activities *shifts from product to process innovations* as the new products or services become more 'standardized' (i.e., 'dominant designs' emerge, cf., Sahal 1981). This 'standardisation' aspect is fostered by the increased costs and diminishing returns inherent to the generation of further product improvements. Consequently, in later phases *price competition* will become a more effective means of competition and emphasis will shift towards improving the related process technologies. The empirical evidence of this 'innovation life cycle' concept has been confirmed both at the sectoral and for the manufacturing sector as a whole (cf.,

Freeman et al 1982, Dosi 1984, Coombs and Kleinknecht 1984, Rothwell and Zegveld 1985).

In this chapter we developed an 'integrating framework' concerning the field of intersection between technological change (innovative activities) and economic transformation. As stated in the introduction, it has not been our purpose to provide a comprehensive overview of all intersections. We have especially concentrated on recent research efforts that consider the life cycles of industries and technologies as *interdependent* (dynamic) phenomena. This line of research has proven to be highly successful concerning the integration of several controversial issues raised in this field. In this respect, we have discussed that the relevance of these issues is not constant over time, but will *change* during the 'swarming' processes. This explains part of these controversies.

In the following chapters, we will return to several items considered in this section and try to integrate these items with the *spatial* dimension of innovative activities. As an introduction to this endeavour, we will in the next chapter turn to the spatial dimension and analyze the ways innovations are dealt with in the spatial innovation literature.

Notes to Chapter 1:

1. This does not imply, however, that there is some kind of a communis opinio regarding the *overall* positive impact of innovations upon the level and quality of all economic variables. Especially concerning the employment variable the argument is still unsettled. Innovations improving the production efficiency, for example, may result in rather large labour displacement effects across sectors. However, in an increasingly (international) competitive environment (cf., Naisbitt 1984), the option of non-innovation might even prove worse.

2. A recent case in point is the social resistance and lack of institutional legislation as regards products generated by DNA-techniques. In this context, technology has progressed much further than the socio-institutional receptivity.

3. Essentially, this is the explanation Freeman et al (1982) provide regarding (long-wave) fluctuations of economic activities.

4. These technological characteristics will largely refer to the manufacturing sector.

5. 'Diffusion' refers here to the diffusion of the new types of products and services among (new) producers of such products and services.

2 Innovations in spatial analysis

2.1 INTRODUCTION

In the foregoing chapter, it has been argued that in economic analysis innovative activities in the production sector are increasingly considered to be an important engine of capitalist competition and economic growth. Contemporary important general trends which can be observed include the transition from hardware to software, manufacturing to services, large scale to small scale production, and hierarchies towards networks (cf., Naisbitt 1984, de Haan and Tordoir 1986 and Orishimo et al 1988).

In the present chapter we will concentrate on the spatial patterns of innovative activities. In accordance with the general 'upswing' in the interest of the role of technological change, the generation and diffusion of innovations in a *spatial context* is nowadays a popular theme in both regional and urban research. In this context, especially innovative activities within the industrial sector have often been studied, although more recently also the importance of these activities within the service sector has been recognized (see also Chapter 9 of the present study).

Essentially, this increased interest in the spatial dimension of innovative activities is brought about by two - mutually interdependent - causes. In the first place at present a common agreement exists concerning the role of innovative activities in determining a firm's competitive performance. At the regional level, this implies that the economic performance of regions or countries is directly linked to the innovative activities performed by firms within the region's boundaries (cf., Molle, 1985).

A second - and related issue - concerns the fundamental *change of regional policy aims* (cf., Ewers and Wettman 1980). In the sixties, one of the main objectives of regional policy was the promotion of the transfer - i.e., re-location - of (high-growth) activities from the congested urban areas to (generally lagging) peripheral areas. For this purpose, regionally varying investment subsidies and a system of constrained and selective granting of investment licences in congested areas were the main policy instruments. The general economic stagnation of the seventies, however, rendered such a policy of relocation largely ineffective as general investments rates and the related mobility of firms declined rather drastically.

Given the relationship between economic performance and innovative activities, the objectives of regional policy have shifted to the promotion of the adoption and generation of *new technologies* in existing firms as well as the generation of *new 'high tech' firms* (by means of 'spin-offs', for example). This shift in emphasis towards the *promotion* of the *indigenous innovation potential* of regions applies to both (previously) assisted and non-assisted areas.

As a consequence of these developments the interest in the *impact* of the regional *'production milieu'* - also labelled *production environment* in this study - upon the innovative performance of firms has increased rather drastically as well. In this context, several regional innovation stimuli - some of which will be discussed in the next section - can be discerned in the literature. The empirical assessment of the impact of these stimuli, however, still remains rather poor up to now (cf., Ewers, 1986).

In spite of this limited empirical verification, several regional authorities bend their instruments towards improving these supposedly favourable stimuli. The explosion in the number of science parks (cf Schwamp, 1987) is a good example in point. Although the impacts on innovative performance (of firms located in such parks) are cumbersome to determine (cf., Moore and Spires, 1983, Nijkamp 1988), there are indications that this effectiveness is inversely related to the total number of science parks in a country (cf., Schwamp 1987, Malecki and Nijkamp 1988).

The increased interest in the spatial dimension of innovative activities is undoubtedly reflected in the great many publications on this topic. In the present chapter we shall make an attempt at identifying a number of general trends. First, in the next section two components of the 'innovative performance' of regions will be discussed, viz., the 'structural' component and the 'production milieu' component. Next, a typology of spatial innovation studies will be developed in section 2.3. A discussion of several types of studies, as well as various (empirical) examples, will be presented in sections 2.4 and 2.5. Finally, section 2.6 contains some concluding and prospective remarks.

2.2 INNOVATIVE PERFORMANCE OF REGIONS

2.2.1 The 'Structural' Component and 'Production Milieu' Component

Essentially, a *spatially* varying pattern of 'innovative output' of firms or establishments - also labelled 'innovativeness' or innovative performance in this study - can be ascribed to two components. The *first* one, viz., the structural component refers to the fact that regions may differ concerning the extent to which its firms are engaged in the field of technological change. In this context, the technological capability of firms in a region will, inter alia, depend on the level of R&D activities performed, the size of the firm and the phase of the 'industry-technology life cycle'. Consequently, the relative innovative performance of regions will depend on their endowment with firms having a high innovation potential on the basis of their *intra-firm characteristics*. In this context the industrial structure of regions, for example, will be important. Hereafter, the endowment of regions with specific *types* of firms - i.e., on the basis of intra-firm characteristics relevant to innovative performance - will be labelled the '*structural*' component.

This 'structural' component also includes the position of firms or establishments in larger networks (cf., Porter 1980, Kamann and Nijkamp 1988). In this respect, the distinction between *firms* and *establishments* - which may be quite irrelevant from a national point of view - is important at the spatial scale of analysis. This observation follows logically from the fact that multi-locational firms often control many spatially dispersed establishments. Consequently, 'in the economic innovation diffusion process the location of the adoption decision and the location of implementation need not be identical' (Pred 1977, pp.124).

So in the (empirical) analysis of the 'structural' component of regions, this distinction between firms and establishments is important. When one is interested in the *spatial* economic impacts of innovations, the establishment level - in which the innovations are implemented - may be preferred (cf., Oakey et al 1980, Thwaites 1982). For practical reasons however, in the sequel both terms will be applied interchangeably so that 'firm' should be read as 'establishment', unless stated otherwise.

The *second* component which will be considered here, refers to the additional impact of *(external) regional stimuli* on the *innovative performance* of firms, i.e., their 'innovativeness'. In the sequel, this component will be labelled the *'production milieu'* impact. In the next section, we will consider several of these (supposed) stimuli.

2.2.2 Relevance of the 'Production Milieu' Component

In the first chapter of this study, several determinants have been discussed which refer to the 'structural' component (e.g., R&D efforts of firms). Determinants referring to the 'production milieu' component will be considered in the next subsections.

In general, in most of the literature no explicit distinction is made between regional stimuli referring especially to the *diffusion* or *generation* aspect of innovations (see sections 2.4 and 2.5). To a large extent this may be due to the fact that these stimuli are often considered to be overlapping, so that no distinct categories can be identified. Consequently, in the following we will not make such a distinction either.

Essentially, the 'production milieu' component may be relevant at two levels. Firstly, this component may be related to the 'structural' component. This would imply that different types of firms - i.e., on the basis of their intra-firm (innovation) characteristics - are attracted to qualitatively different types of regions.

Secondly, the 'production milieu' component may influence-ceteris paribus the intra-firm characteristics or 'structural' component - the innovative performance of firms. This would imply that the innovativeness of *identical* firms - as far as their intra-firm characteristics are concerned - *varies* according to their *location in space*. As stated above, this 'latter' effect will be denoted as the '*production milieu*' impact.

2.2.3 Four Clusters of 'ProductionMilieu' Variables

In the following analysis of key factors of the 'production milieu' we will discuss 4 clusters of production environment variables which are often expected to affect the innovativeness of (existing) firms. Also, these variables are often assumed to be important when considering incubation areas for the take off of new life cycles of industries and technologies (see the next chapter, for example).

(A) Agglomeration of (Different Types) of Firms

In spatial innovation research regional variations in the *spatial concentration* of firms and/or households is often considered to be important. In this context, a distinction is often made between '*localization*' and '*urbanization*' economies.

> Localization economies occur when many plants in a single industry
> acquire various cost savings - e.g., lower per unit input or
> service expenses - by clustering in the same urban area' (Pred
> 1977, pp.96).

This spatial clustering of *similar* firms may also imply a higher incentive to innovate in order to attain a share of the market (cf., Hansen 1983 and Mouwen 1984).

On the other hand, 'urbanization' economies, e.g., lower costs of information acquisition, result from the enlargement of the total economic size of all industries in a single urban area' (Pred 1977, pp.96). The spatial clustering of *dissimilar* firms and institutions (for example, public R&D institutions) may affect the outcome of the innovation process - i.e., 'innovativeness' - of individual firms in a positive way. As innovations (especially in the introduction phase) have to be changed frequently, a high diversity of suppliers and buyers close at hand may reduce the risks involved in this process of change and, consequently, increase the number of innovations being 'generated' and adopted (cf Hoover and Vernon 1959, Vernon 1960, Thwaites 1978, Carlino 1978, Brown 1981). In this context, the (supposed) importance of *'face to face'* contacts for innovative activities has to be mentioned also (cf., Thorngren 1970).

Clearly, regarding these 'spatial economies of scale' metropolitan or central areas are generally expected to be in a favourable position (cf., Batten 1982, Lambooy 1984, Kok et al 1985, Moss 1985). It should be added, however, that according to some authors there may also be certain limits concerning these spatial economies (cf., Nijkamp and Schubert 1983, Mouwen 1984 and Camagni and Diappa 1985). In this interpretation, *diseconomies* may come to the fore in case the spatial concentration of firms and/or population becomes too large compared to the 'carrying capacity' of the area. This will result in 'congestion' effects.

(B) Demography and Population Structure

The population base - or *'market area'* - of a region is often considered to be positively related to the generation or adoption rate of innovations in individual firms (cf Berry 1972, Pred 1977, Brown 1981). This applies especially to those innovations which need a high market potential in order to be profitably implemented (i.e., 'high order' goods in the framework of Christaller 1933). Also the spatial dispersion of *special subgroups* - e.g., of 'opinion leaders' or 'minority groups' - is often considered to be important (cf., Berry 1972, Pred 1977 and de Ruijter 1983).

However, especially the spatial dispersion of specific scarce *labour forces* - i.e., technical and managerial personnel - is usually considered to be important. This personnel appears to be attracted to locations with a rich variety of *cultural and educational amenities*, i.e., in or near the larger metropolitan areas. Consequently, the location of R&D facilities - whether private or public - is to a large extent determined by the locational preferences of these employees (cf., Malecki 1979a, 1979b, Thwaites 1982, Bushwell 1983, De Jong and Lambooy 1984, Andersson and Johansson 1984, Oakey 1984 and Johansson and Nijkamp 1987). This pattern is a clear example of the fact that the 'production milieu' and 'structural' component of regions may be interwoven.

(C) Information Infrastructure

Availability of information is generally regarded as an important stimulus to the generation and adoption of innovations of firms (cf., Feller 1973, 1979, Pred 1977, Norton 1979, Oakey 1984 and Gibbs and Thwaites 1985). Especially Pred (1977) can be credited for having demonstrated that the availability of specialized information is spatially biased and oriented towards the larger metropolitan areas.

> The spatial biases of innovation-relevant information circulation are normally such that any growth-inducing innovation perfected or introduced within the city-system at one major metropolitan complex is more likely to be adopted early on a larger or multiple basis at other major complexes than at smaller places within the system (Pred 1977, pp.179).

In this context, the spatial dispersion of *public research institutes, universities, institutes of technology* and *knowledge transfer centres* favours the availability of (technical) information in the larger metropolitan complexes (cf., Malecki 1979a, 1980 and Nijkamp 1988). In this respect, educational and training facililities are increasingly recognized as important vehicles for the technological capabilities of a region (cf., Malecki and Nijkamp 1988).

The *'spin-off'* phenomenon has to be mentioned also, as people working in such research institutions may often 'spin-off' and establish a new (innovative) firm (cf., van der Meer and van Tilburg 1983, Oakey 1984 and Rothwell and Zegveld 1985). In many cases such firms will be located near the 'parent' institution because of subcontracting arrangements with this institution and the tendency of new 'entrants' to locate their new firm near their place of residence (cf., Gudgin 1978 and Aydalot 1984).

(D) Physical and Institutional Infrastructure

The availability of public infrastructure is often perceived as a necessity to the generation or adoption of innovations:

> the availability of a satisfactory level of public infrastructure capital stock (in its broadest sense) shapes the necessary conditions for innovative capacities in an area (Nijkamp 1983, pp.80).

In this context, availability of rapid transportation networks (e.g., airports) are often considered to be especially important to 'high tech' sectors (cf., de Jong and Lambooy 1984, Gibbs and Thwaites 1985, Camagni and Rabellotti 1986, Button 1988). At present, also the availability and quality of *telecommunication networks* in a region may be an important prerequisite for innovative activities in the field of computer related technologies (cf., Brown 1981, Freeman 1987a).

The *financing* of innovative activities is often problematic given the high risks involved (cf., Oakey 1984 and Rothwell and Zegveld 1985). Therefore, spatial variations in the availability of *'venture capital'* is often considered important to local innovative activities (cf., Lambooy 1978, Bushwell 1983, Oakey 1984 and Stöhr 1985). Oakey (1984), for example, stresses the positive impact of venture capital availability on innovative activities in the Bay Area in the U.S. as compared to Scotland and South East England. Stöhr (1985), on the other hand, illustrates the role of venture capital in relation to the Mondragon project in Spain.

As a consequence of the (supposed) impact of such 'production milieu' variables as described above, considerable research efforts have been devoted to 'approximating' several of these variables. In the Netherlands, for example, Vlessert and Bartels (1985) and Mouwen and Nijkamp (1985) have analyzed the spatial dispersion of (employment in) *knowledge centres*. Van der Meer and Brand (1987) present several *'production milieu' indicators* for 26 (large) cities in the Netherlands. The Netherlands Economic Institute (1984), on the other hand, has constructed a large number of such indicators for each COROP-region[1] in the Netherlands. We will return to the NEI study in Chapter 6 of this book.

In this section we have discussed several variables of the 'production milieu' which are often considered to be conducive to the innovative performance of firms. As stated above, the 'production milieu' may be relevant in two respects. *Firstly*, the characteristics of this 'milieu' may be interwoven with the internal characteristics - i.e., the 'structural' component - of local firms. This notion is in accordance with conceptualizing the organization of firms as open systems (cf., Ewers and Wettman 1980). In this notion, the internal characteristics of firms will be dependent on their 'task environment'.

This 'task environment', on the other hand, is again related to the space-time limitations of the locality in which a specific firm is resided. Consequently, the internal characteristics of firms will be related to the features of their local environment. An example of such a relation has been presented above where we discussed how firms performing R&D activities may be 'forced' to locate near large-amenity-rich - metropolitan areas because of the locational preferences of their R&D employees.

Secondly, the 'production milieu' component may be relevant as an independent determinant of the adoption or 'generation' rate of innovations at the individual firm level. So this 'production milieu' effect refers to the additive impact of regional stimuli - ceteris paribus the intra-firm characteristics - on the innovativeness of firms.

2.2.4 Formalization of both Components

In formal terms the spatial analysis of innovative activities can be presented as follows:

$$ECON_r = f(INN_r, O_r), \quad f^1 > 0 \qquad (1)$$

in which:

$ECON_r$ = economic performance of firms located in region r

INN_r = innovative performance - or 'innovativeness' - of these firms

O_r = other - i.e., non-innovative - characteristics of firms (e.g., educational level of labour force, management capabilities, and the like)

This equation reflects the above mentioned trend in which the economic competitiveness of (firms in) regions is increasingly acknowledged to depend inter alia on the innovative activities performed. On the other hand, spatial variations in innovative performance (of firms) can be considered to be related to the two components discussed above, viz., the 'structural' and 'production milieu' components:

$$INN_r = g(STRUC_r, PM_r) \qquad\qquad (2)$$

in which:

$STRUC_r$ = 'structural' component of firms in region r

PM_r = 'production milieu' component in region r

Next, the 'structural' component can be presented as follows:

$$STRUC_r = h(IP_{0r}, IP_{1r}, \ldots) \qquad\qquad (3)$$

in which:

IP_{ir} = (level of) *intra-firm* innovation 'inputs' Pi (i = 0,1,...)
    of firms located in region r

So according to the 'structural' component, regions may differ regarding the innovative performance of their firms because of spatial variations in the (level of) *intra-firm innovation inputs* (e.g., R&D efforts). As stated above, the configuration of intra-firm characteristics in a region may depend on the characteristics and qualities of the given regional 'production milieu'. Empirical examples of the measurement of spatial variations in intra-firm innovation input efforts - i.e., *'innovation potential'* - will be presented in section 2.5 and subsequently in the second part of this study.

Next, the 'production milieu' component can be presented as follows:

$$PM_r = i(A_r, B_r, \ldots) \qquad\qquad (4)$$

in which:

A_r, B_r = value of 'production milieu' characteristics A, B, ... in
    region r.

The determinants of the 'production milieu' component refer to the *additional* impact - i.e., besides the impact of intra-firm innovative efforts - of *external* regional stimuli on the 'innovativeness' of firms.

So in spatial innovation research the analysis of 'production milieu' and/or 'structural' aspects of regions with respect to innovative activities performed is often a central theme. The distinction between 'structural' and 'production milieu' aspects might serve here as a first classification criterion for discriminating between different types of studies. However, as will be indicated later on in this chapter, several spatial innovation studies display a 'mixed' character in this respect and classification according to a 'production milieu' or 'structural' angle would be rather arbitrary. Consequently, hereafter we will not use this criterion in a strict sense, but rather indicate which aspect is especially relevant in certain (types of) spatial innovation studies.

After this concise discussion of the 'production milieu' and 'structural' component, in the next section our framework for classifying spatial innovation studies will be considered.

2.3 A TYPOLOGY OF SPATIAL INNOVATION STUDIES

In this section we will develop our classification scheme for distinguishing various types of spatial innovation analyses. This typology will serve as a framework for considering various types of approaches in spatial innovation analysis as well as for selecting basic research 'gaps' to be taken up later on in this study. In this respect, our typology of spatial innovation studies will be based upon the following classification criteria:

I. The Types of Innovative Activities Studied

In this respect, a distinction will be made between the *generation* and *diffusion* (cf., Stoneman 1983) of innovations. Historically, especially the latter theme has been studied in spatial analysis (cf., Brown 1981). In this respect, spatial *diffusion* research has not been limited to the analysis of techno-economic - or growth inducing-innovations, but has in the past also considered the spatial diffusion of cultural patterns (Sauer 1952), agricultural innovations (Hägerstrand 1967), classical town names (Zelinski 1967) and information. Only more recently, the spatial diffusion of techno-economic innovations has gained more interest.

However, given the acknowledged importance of the *indigenous generation* of innovations within firms as an effective means of gaining a competitive edge, the main emphasis in spatial innovation analysis seems to have shifted towards this 'generation' issue. We will return to the above mentioned issues in sections 2.4 and 2.5.

II. The Sectors in which Innovative Activities are Studied

This classification criterion refers to the question whether technological change is studied in the context of *existing* ('traditional') or relatively *new* ('high tech') firms and/or sectors.[2] As will be shown in the next section, especially the latter aspect is quite popular nowadays. In this respect, the *urban incubation hypothesis* (cf., Hoover and Vernon 1959, Jacobs 1966) - nowadays increasingly interpreted in terms of the role of metropolitan areas as 'breeding places' for new *innovative* firms (cf., Andersson and Johansson 1984, van der Knaap and Wever 1987, Davelaar and Nijkamp 1987a, 1989g) - is again gaining interest. In subsequent chapters of this study, the incubation potential of metropolitan areas as regards the 'take off' of new life cycles of industries and technologies will be reconsidered.

Figure 2.1 is a summary of our framework for classifying spatial innovation studies.

Figure 2.1 A Classification Scheme for Spatial Innovation Studies

Innovation Sectors	Generation	Diffusion
New ('high tech')	section 2.5.2	section 2.4.3
Existing ('old')	section 2.5.3	section 2.4.4

In the next sections various types of studies - categorized according to Figure 2.1 above - will be discussed. We will start with *diffusion* oriented studies.

2.4 DIFFUSION OF INNOVATIONS AS AN OBJECT OF SPATIAL ANALYSIS

2.4.1 Diffusion Analysis

In (spatial) innovation diffusion research a generally accepted concept is the S-shaped diffusion curve (cf., Griliches 1957, Davies 1979, Metcalfe 1981, Andersson and Johansson 1984, McArthur 1987). According to this concept, the cumulative level of adoption over time proceeds according to an S-shaped curve as sketched in Figure 2.2 below.

Figure 2.2   The Adoption Curve

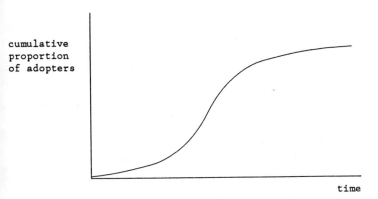

Essentially, one of the main purposes of diffusion analysis is the description and causal explanation of variations in *adoption time* (i.e., date of first adoption) - or *adoption rate*[3] - between households, firms, regions, and the like. In terms of Figure 2.2 above, the purpose of a causal analysis is the identification of those factors which determine whether a household, firm, or region will be located in the left or right part of this figure - i.e., whether it will be an early or late adopter, respectively. In industry-specific diffusion studies, for example, such explanatory variables relate to size of the firm, market structure, profitability of the innovation (cf., Nabseth and Ray 1974, Davies 1979 and Kamien and Schwartz 1982).

One of the basic problems inherent in such an explanatory analysis appears to be the identification of the potential set of adopters and the exact date of first adoption (cf., Davies 1979, Batten and Johansson 1987). In case of lack of such information the adoption rate - at a certain moment of time - cannot be determined.

2.4.2 Spatial Diffusion Analysis

The *spatial* diffusion of innovations is considered to be a valid topic for geographical analysis, as

> The spread of a phenomenon, idea or technique throughout a
> population or region incorporates basic geographic elements of
> distance, direction and spatial variation, and thus forms a valid
> field of geographic interest (Brown 1981, pp.16).

As will be clear, spatial diffusion analysis concentrates in particular on a description - and causal explanation - of spatial variations in adoption-time or adoption-rate.

Especially Hägerstrand (1967) is often credited for being one of the first geographers to analyze the spatial diffusion process of innovations. However, this topic has also been studied in earlier work of - especially - cultural geographers (cf., Brown 1981). Hägerstrand (1967), however, provided a more in-depth analysis of the *causes* of spatially variant diffusion processes. In this context, he considered the adoption of an innovation as resulting from a *learning and communication process*. By means of this process, a potential adopter learns about the existence and characteristics of a particular innovation. This interpretation implies

> that factors related to the effective flow of information are most
> critical and, therefore, that a fundamental step in examining the
> process of diffusion is identification of the spatial
> characteristics of information flows and resistances to adoption
> (Brown 1981, pp.19).

As described in the foregoing section, especially Pred (1977) has demonstrated that these flows of specialized information are spatially biased towards the larger metropolitan complexes. This line of research especially stresses the behaviour of (potential) adopters - i.e., the demand aspect of the diffusion of an innovation. Brown (1981) labels this the *'adoption perspective'* of innovation diffusion. In this

analysis, it is implicitly assumed that every potential adopter has an
equal 'opportunity' to adopt and *spatial variations* in adoption depend
on *spatially biased information flows.*

As Brown (1981) convincingly states, however, the role of the
supply side - i.e., the 'market or infrastructure perspective' - should
be considered also. This perspective 'takes the stance that the
opportunity to adopt is egregiously and in many cases purposely unequal.
Accordingly, focus is upon the process by which innovations and the
conditions for adoption are made available' (Brown 1981, pp.7).

Historically, spatial diffusion analysis largely ignored the
growth-inducing (techno-economic) innovations we are interested in. So
even in 1980 Ewers and Wettman observed that: 'the question of whether
there are locational differences in the innovation performance of firms
has been given only little attention by innovation and diffusion
research up until now' (Ewers and Wettman 1980, pp.166). Besides and
above this,

> the small number of empirical studies attempting to link diffusion
> with the spatial spread of economic growth have centred mostly on
> innovations which are artifacts of growth, such as TV-ownership,
> rather than on true growth-inducing innovations such as new
> products and services, new production and communications
> technology, and new ways of performing or structuring the
> operations of business and government organizations (Pred 1977,
> pp.124).

However, given the more recently acknowledged importance of
techno-economic innovations for (regional) economic growth perspectives,
such innovation studies are gaining interest nowadays (see in particular
subsections 2.4.3 and 2.4.4).
Concerning these innovations, two broad categories of spatial *diffusion*
analyses can be distinguished (cf., our innovation typology developed in
section 2.3). In the following pages, we will present a general
discussion of both types and discuss some relevant examples. We will
start with the right upper part of Figure 2.1.

2.4.3 Spatial Diffusion of New Innovative Economic Activities

In this type of research, the main emphasis is on the analysis of the space-time trajectory of new innovative firms, i.e., of an 'entrepreneurial' innovation (see below; see also Chapter 9 of this study). From a theoretical point of view this trajectory is often expected to start in metropolitan areas (see also the next chapter). As stated above, the urban incubator hypothesis (cf., Hoover and Vernon 1959, Jacobs 1966, Davelaar and Nijkamp 1986, 1987a, 1988) is increasingly interpreted as referring to new *innovative* firms nowadays.[4] Consequently, at the *regional* level metropolitan areas - though not necessarily the urban cores themselves (cf., Scott 1982, Storper 1986)- are expected to be important *initial* breeding places of new economic activities.

The 'explanations' presented for such a spatial pattern of diffusion again refer to 'the general properties of space' as discussed in section 2.2.2, i.e., distance and 'the varrying degree of spatial concentration of human activities between different locations' (Ewers and Wettman 1980, pp.169). In this context, Berry (1972) also mentions the possibility of a 'market searching' process in which the newly created market opportunities (resulting from such 'entrepreneurial' innovations) are exploited on the basis of a larger-to-smaller sequence. Brown (1981), however, states that such a process might especially be relevant in the context of 'profit-motivated diffusion under a *centralized* decision-making structure'. Concerning a *decentralized* decision structure

> the profitability associated with a particular urban center will be important, but should operate primarily as a threshold for agency establishment rather than as a basis for ranking alternative locations (Brown 1981, pp.169).

In this case, however, *information linkages* among places may be important in determining the actual ranking of those places which exceed the critical market threshold (cf., Pred 1977 and Brown 1981). Besides and above this

> It seems reasonable to hypothesize that higher-order innovations feature a higher elasticity of agency profitability with regard to urban area sales potential, and therefore a greater impetus towards hierarchically ordered locational priorities (Brown 1981, pp.62).

So in spatial analysis the diffusion of such new innovative types of firms[5] - or 'entrepreneurial innovations' (cf., Berry 1972, Bearse 1978) - is generally conceptualized as being determined by three interdependent 'forces' (see also Pedersen 1970, 1975, Berry 1972, Brown 1981)[6], viz.,:

a. urban size or rank order of a certain metropolitan area: the so-called 'hierarchical effect'
b. relative location of a metropolitan area: the so-called 'neighborhood' or 'contagion' effect
c. minimum market threshold size

Pedersen (1970, 1975), Berry (1972) and Brown (1981), for example, integrate these 'forces' in a model - determining the adoption time of new 'entrepreneurial innovations' in a region - in the following way. First, the amount of information (I) - related to the new 'entrepreneurial innovation' - reaching a certain area i in period t must exceed a critical market threshold size (F). The 'amount' of information reaching a region i is determined in the following way:

$$I_i(t) = \sum_{j=1}^{n} K * \left[\frac{P_i * P_j(t)}{(d_{ij})^x} \right] > F \qquad (5)$$

in which:

j : metropolitan areas in which the new 'entrepreneurial innovation' has already been established at period t

P_i, P_j : population size of i and j, respectively

d_{ij} : distance separating i and j

K, x : scale parameters

Secondly, a (fixed) 'minimum market threshold' (G) condition can be incorporated in the following way:

$$P_i > G \qquad (6)$$

implying that diffusion does not proceed below the critical 'urban threshold' size G.

Thirdly, a probabilistic mechanism can be introduced in the following way:

$$S_i = 1 - e^{-(P_i *q)} \qquad (7)$$

in which,

S_i = probability that - in case conditions (5) and (6) are
       fulfilled - the new 'entrepreneurial innovation
       will indeed be taken up by local entrepreneurs
       and result in a new agency establishment in region i

$P_i *q$ = total number of (potential) entrepreneurs in region i
       (taken to be a fixed fraction q of total population
       residing in region i)

It is easy to see that the above mentioned forces (a) - (c) have been incorporated in the first two conditions as determinants of the *adoption-time* in a set of regions.

After a region has 'received' the new 'entrepreneurial innovation', a subsequent analysis typically concentrates especially on the *demand* side or 'adoption perspective' of innovation diffusion. In this context, the adoption - or *use* of the new 'entrepreneurial innovation' (Berry 1972 labels this a 'household innovation') within 'infected' regions - is often perceived as proceeding in a contagious pattern from the center of such regions onwards[7]: 'a logistic spatial-temporal trend may be postulated for the utilization of the innovation by households within its field of influence' (Berry 1972, pp.118).

Consequently, entrances of *other* new innovative entrepreneurs and (the spatial consequences of) *competition* between such entrepreneurs on the basis of *further improvements* in both the (characteristics of the) new types of products or services supplied and the related process technologies are generally ignored in this type of analysis.[8] So in this type of analysis, technological change (on the supply side) is only considered in as far as an 'entrepreneurial innovation' will be made available or not in a specific region at a certain moment of time.

Empirical studies which analyze the (complete) space-time trajectory of new innovative sectors are relatively scarce, however. This may be understood from the fact that such an analysis would require extensive *time-series* data concerning the *spatial* evolution - i.e., birth, growth and decline - of such specific new activities (see also

Molle and Vianen 1985, concerning micro-electronic firms). As will be discussed more thoroughly in Chapter 9 of this study, such data are generally lacking. Consequently, this trajectory is often cut off by means of the analysis of the spatial distribution of such new activities at a few moments in time, rather than presenting a 'complete' spatial life cycle analysis.

Some empirical examples in this respect are: the space-time trajectory of TV-stations in the U.S. (Berry, 1972), the analysis of the producer service sector in New York (Bearse 1978), computer-service and software firms in the Netherlands (Koerhuis and Cnossen 1982), the consultancy sector in Sweden (Andersson and Johansson 1984), and selected producer service subsectors in the Netherlands (Davelaar and Nijkamp 1989a, 1989b). The empirical findings in these studies largely support the theoretical framework sketched above (see also Chapter 9 of this study). In initial phases of the space-time trajectory of such innovative sectors a (strong) metropolitan bias can be disentangled.

Andersson and Johansson (1984), for example, show that in 1970 71.5% of total employment in the consultancy sector in Sweden was located in metropolitan regions versus 64.4 % in 1980 (indicating a spatial decentralization of this sector). The same conclusion applies, mutatis mutandis, to the other examples mentioned. In Chapter 9 of this study, the incubation potential of Dutch metropolitan areas with respect to various highly dynamic producer service (sub)sectors - linked to the more recent information 'technology system' - will be analyzed more in depth.

2.4.4 Spatial Diffusion of New Technologies in Existing Firms

The second type of diffusion research which will be discussed in this section concentrates on the analysis of the space-time trajectory of *specific new technologies*, which are to a large extent implemented in *existing* firms or sectors. This type of research can be placed in the right lower part of Figure 2.1. Typically, this type of research considers spatial variations - in adoption time and/or adoption rate- of some pre-specified and supposedly growth-inducing innovations among an existing set of firms. Recent examples of such innovations refer to CNC-machines, CAD/CAM systems, computers, micro-processors, robotics.

Clearly, spatial variations in adoption may be due to (a combination of) 'structural' and 'production milieu' impacts. Ewers and Wettman (1980) label this latter impact the 'regional factor' in diffusion - i.e., the regional effect after compensating for the 'structural' (intra-firm) characteristics of the pool of firms located in a region. Clearly, the 'structural' component will often be highly relevant in determining regional variations in adoption-time (or rate) of new technologies.[9]

However, the determination of the impact of the (additional) 'production milieu' impact is often troublesome. Such an effect can only be established after 'compensating' for the impact of the 'structural' component, i.e., the intra-firm characteristics. The number and quality of the data available often preclude such an in-depth analysis, however.

In this context, the empirical validation of both impacts mentioned above is also seriously hampered by the fact that (see subsection 2.4.1) it is often difficult to determine à priori the potential applications - i.e., the set of potential adopters - of broadly applicable new technologies. In this respect, two strategies can be pursued:

1. To select a specific new technology which is expected to be relevant to a specific group of firms only (e.g., electric furnaces in the steel industry in Canada; cf., Martin et al 1979)

2. To study the space-time trajectory of broadly applicable innovations in the context of specific sectors only (e.g., the adoption of CNC, CAD/CAM and micro-processors in 9 metal working industries in the UK; see Thwaites 1982, Alderman 1985).

Theoretically, the impact of the 'production milieu' component on the timing and level of adoption of such technologies - ceteris paribus the 'structural' component - is generally expected to be determined by the 'forces' mentioned in section 2.4.3 above. These forces would favour a constellation of 'hierarchical' and 'neighborhood' patterns of diffusion of such technologies (cf., Thwaites 1978, Ewers and Wettman 1980).

Empirical analyses tend to confirm the initial leading position of (firms located in the) larger metropolitan complexes (cf., Martin et al 1979, Ewers et al 1979, Thwaites 1982, Camagni 1984, Northcott et al 1984, Alderman 1985). However, the basic question whether this metropolitan bias results from 'structural' and/or 'production milieu' impacts, generally remains unresolved. (This is often due to the low number of observations).

Alderman (1985), for example, applies a Davies' type of approach (cf., Davies 1979) to the diffusion of NC, CNC machines and computers among 9 metal working industries in the UK. He demonstrated that the diffusion of these technologies lags considerably behind in establishments located in the Northern Region (compared to those located in the North West and South East). However, it is also acknowledged that 'Such variations are likely to reflect the different characteristics of the stock of establishments in each region' (Alderman 1985, pp.12). As revealed in his logit-analysis, such (intra-establishments) characteristics might refer to employment size of the establishment, availability of on site R&D, type of sector, prior investments, diversity of operations performed and status of the establishment.

2.4.5 Pros and Cons of Spatial Diffusion Research

In concluding this section, we can state that spatial diffusion research has especially concentrated on the (causal) analysis of the space-time trajectory of (supposedly) *given* and *constant* innovations and innovative activities. In this respect, a distinction can be made between the spatial diffusion of 'entrepreneurial' innovations - i.e., new ('Schumpeterian') types of firms/sectors - and the diffusion of *specific new technologies* among *existing* firms.

The identification and explanation of spatial variations - in both adoption-time as well as adoption-rate - are the main purposes of analysis. Especially in the case of the spatial diffusion of specific new technologies both 'structural' and 'production milieu' impacts are generally considered to be relevant. Theoretically, the 'production milieu' impact - in both types of diffusion analysis discussed above- is generally considered to favour a (constellation of) hierarchical and contagious patterns of diffusion.

For the purposes of our study, especially the emphasis on the *dynamic* aspects of *technological change* in a *spatial* context is highly relevant. However, two important (interrelated) shortcomings are also noteworthy:

a. The characteristics of the innovations - or new innovative firms-studied remain the same in the course of the diffusion process. However, in the previous chapter it has been discussed that diffusion is generally a '*creative*' process characterized by *further technological improvements* incorporated within new and existing firms. These improvements are an important means of competition between firms. In *spatial diffusion* analysis however - i.e., the first type of diffusion studies discussed above - competition between new innovative firms (and regions) on the basis of further innovative activities[10] are generally left aside. In this way the importance - and spatial impacts of - the *endogenous generation* of further innovations during the 'swarming' processes (see the previous chapter) are largely ignored.

b. Also and accordingly, at the theoretical level spatial variations in the *economic* impact of innovations are not so much influenced by spatial variations in indigenous innovative activities, but rather by (a constellation of) 'a-technological' conditions and assumptions such as:

- spatial variations in the *time-lag* of *adoption* of exogenously
  generated ('entrepreneurial') innovations
- a *minimum urban threshold* criterion below which diffusion does not
  proceed
- the assumption that 'the income effect of a given innovation is a
  declining function of time' (Berry 1972, pp.108)
- the assumption that *wage rates* related to such innovations
  decrease over time (cf., Thompson 1968).

Consequently, in spatial diffusion research the role of technological change is limited to an 'exogenous' generator of specific once and for all 'static' ('entrepreneurial') innovations, the diffusion of which is studied in a space-time framework. The central theme in diffusion analysis is the description and explanation of regional variations in the *timing* and/or *adoption rate* of these innovations as both are considered to be related to economic performance.

Consequently, regional *'indigenous'* possibilities to *generate* qualitative different types of innovations, are largely ignored in diffusion analysis. The ignorance of the *regional 'indigenous'* and *'creative'* aspects of diffusion (see the foregoing chapter) - i.e., the *endogenous generation* of further innovations as a means of competition between firms and regions - confines the use of this type of analysis rather seriously. In the next section, we will discuss spatial innovation studies which especially stress the *'generative'* aspects of innovative activities.

2.5 THE GENERATION OF INNOVATIONS IN A SPATIAL CONTEXT

2.5.1 Endogenous Generation of Innovations

In the preceding chapter we have discussed the role of technological change as an important 'engine' of competition and economic growth. In the foregoing sections, it has been demonstrated that this point of view also has its spatial counterpart. The indigenous innovation potential of *all* regions - both assisted and non-assisted- to take part in the *generation* of further technological advances is increasingly considered crucial for regional economic growth nowadays. So in line with this 'Schumpeterian view', the number of studies analyzing the regional *'indigenous'* capabilities to *'generate'* (techno-economic) innovations, has increased rather drastically in the eighties.

Contrary to spatial *diffusion* research, these 'generation' studies do *not* concentrate on the spatial dispersion of *specific* exogenous and supposedly constant (entrepreneurial) innovations. On the contrary, in these studies technological change is (implicitly) considered to be a multi-farious - *non-exogenous* - phenomenon which can be manipulated by means of input efforts (R&D) devoted to it.

Consequently, these 'generation' studies especially concentrate on the regional indigenous potential to generate qualitative different types of innovations. The parallel with economic analysis discussed in the foregoing chapter - i.e., the shift in conceptualization from 'manna from heaven' to technological change as an effective means of competition - will be clear.

As will be demonstrated below, these studies are especially biased towards the spatial dimension of innovative efforts taking place within relatively *'new' (or 'high-tech') sectors* (i.e., the upper left part of our innovation typology in Figure 2.1). The reason for this bias, can be traced back to the expectation that the more significant technological advances will be especially generated within these sectors. Consequently, we will first discuss - and present several examples of- regional innovation studies which relate to the *upper left* part of our innovation typology.

2.5.2 Innovative Activities in 'New' ('High Tech') Sectors

In this type of analysis the following approaches can be distinguished:
a. *Sector* studies
b. *Innovation input* studies
c. *Innovation output* studies

These approaches will (briefly) be considered below.

2.5.2.1 Sector Studies

The purpose of the 'sector studies' is to analyze the spatial dispersion of employment or establishments of a priori selected 'high tech' sectors. This dispersion is often considered to be a proxy for the indigenous technological capabilities. In general, 'high tech' selection is based upon SIC-codes of firms. The 'translation' from 'high tech' to SIC-codes is often troublesome, however. In this respect, problems arise both from the apparent lack of an adequate definition of what 'high tech' implies as well as inadequacies inherent to SIC-codes in general (cf., Alders and de Ruijter 1984, Bouwman et al 1985a, 1985b, Markusen et al 1986).

Consequently, in this type of analysis, several (different) pragmatic solutions are often applied, resulting in considerable variations regarding the final selection of 'high tech' sectors. Also concerning the *size* categories of selected firms these studies may differ considerable. However, the 'overall bias' in these studies appears to be related to the *larger* establishments (for the same reasons as the bias towards 'high tech' sectors mentioned above).

Some (empirical) examples of this type of analysis are OTA (1984), de Jong (1984), Alders and de Ruijter (1984), NEI (1984), Bouwman et al (1985a, 1985b), Molle and Vianen (1985), Keeble and Kelly (1986), Markusen et al (1986), Keeble (1986), de Ruijter et al (1986), Leigh et al (1986), Louter (1987) and Machielse and de Ruijter (1989). By comparing the spatial concentration of selected 'high tech' sectors with, for example, the spatial dispersion of total (manufacturing) employment - e.g., via a location-quotient method - the (relative) indigenous innovation potential of regions is often gauged.

In this respect, a 'general' finding of these studies appears to be that in such *relative* terms the selected sectors appear to be biased towards the larger metropolitan or central areas. Alders and de Ruijter (1984) demonstrate that - as far as the Netherlands is concerned - this statement applies especially to advanced types of *service* sectors. As regards the Dutch *manufacturing* sector, Bouwman et al (1985a) select 10 'high-tech' manufacturing subsectors (at a 4-digit SIC level). The spatial dispersion of these 'high tech' establishments - i.e., at the regional level of Dutch provinces - compared to the *total* number of manufacturing establishments is depicted in Table 2.1 below.

Table 2.1   Spatial Dispersion of 'High-Tech' Establishments in 1984

Province	Number of 'high tech' establ.	Number of manu-facturing establ.	Location quotient
Groningen	36	1,499	0.88
Friesland	26	1,860	0.55
Drente	26	1,005	0.90
Overijssel	76	3,365	0.69
Gelderland	119	5,748	0.74
Utrecht	129	2,902	1.59
Noord-Holland	268	8,405	1.15
Zuid-Holland	340	9,220	1.33
Zeeland	22	1,026	0.77
Noord-Brabant	159	7,579	0.76
Limburg	73	3,545	0.75
ZIJP	13	253	2.00
Total	1,287	46,407	

Source: Bouwman et al (1985a), pp.38

From this table it can be concluded that the provinces in the Western (i.e., central) part of the Netherlands - i.e., Utrecht, Noord-Holland and Zuid-Holland - perform relatively favourable.

Looking more specifically at the *intra*-metropolitan level, the favoured 'high tech' locations nowadays often appear to be *suburban* in character and outside the old industrial areas (cf., Scott 1982, De Jong and Lambooy 1984, Aydalot 1984, Camagni and Rabellotti 1986, and Storper 1986). In the course of time, most 'high tech' sectors display a pattern of spatial *deconcentration*[11] (cf., Markusen et al 1986, Barkley 1988).

In view of our discussion in section 2.2 however, some serious methodological problems are involved in 'approximating' the regional indigenous innovation potential by means of this type of analysis:

- It is (implicitly) assumed that the intra-firm characteristics of selected 'high tech' firms - which act upon the innovative capability of these firms - do not vary over space. Or, stated otherwise: to which degree can spatial dispersion of 'high tech' *employment* or *establishments* be considered as an adequate proxy for *innovative* efforts?
- Secondly, in the absence of innovation *output* data, the impact of the (additional) 'production milieu' impact - see again subsection 2.2.4 equation (2) - cannot be determined and is ignored in these studies.
- Spatial dispersion of innovative activities within *non-*'high-tech' firms may differ from 'high tech' firms.
- The selection of 'high tech' sectors is not unequivocal.

'Advantages' of this type of analysis refer to the '*overall*' picture, (i.e., of *all* selected high tech subclasses[12]) these studies are often able to present. However, given the above-mentioned methodological considerations, one should be cautious in interpreting the regional dispersion of 'high tech' employment (or establishments) as proxies for regional innovative performance. The ignorance of, for example, spatial variations in *types* of 'high tech' firms as well as the *impact* of the '*production milieu*' component on innovative performance of these firms hampers such a extrapolation. Concerning these issues micro-oriented approaches may be more fruitful. We shall subsequently discuss these approaches.

2.5.2.2 Innovation Input Studies

This type of analysis examines spatial variations in intra-firm *input* (or innovation potential) characteristics - i.e., the 'structural' component - considered to be important for the innovative performance of firms. In this respect, an important intra-firm characteristic relates to the *R&D efforts* - in monetary or employment terms - devoted to innovative activities. Malecki (1979a), for example, states: 'R&D is a necessity for firms to be innovators and successful adopters, and through the efforts of these firms, the regions in which they are located maintain a comparative advantage in technology, new products, and new industries' (pp.322).

Consequently, spatial variations in the 'structural' component are generally 'approximated' by means of spatial variations in the *R&D efforts* of individual firms. Although the importance of R&D inputs to the generation of innovations is generally acknowledged, 'the study of R&D activity in a spatial context has been limited' (Howells 1984, pp.14).[13]

Just like the 'sector studies', also in this type of analysis the 'production milieu' impact is generally ignored. So the possibility of regional variations in the '*productivity*' of intra-firm innovation *input* efforts (like R&D) - caused by location in qualitatively different production environments - is generally ignored.

Concerning the spatial dimension of R&D activities it is useful to discriminate between the following types of studies:

1. '*Aggregate*' analyses which study the spatial distribution of *total* R&D activities - approximated by means of the spatial dispersion of *laboratories* or *R&D employees*. Typically, linking these activities to individual firm efforts is not possible (cf., Bushwell and Lewis 1970, Malecki 1979a, Howells 1984). Concerning R&D *employment*, data limitations may even preclude a differentiation according to *private* and *public* R&D efforts.

In these aggregate analyses, (total) R&D efforts appear to be biased towards the more central and urbanized areas. Malecki (1979a) for example - in analyzing the location of total R&D efforts at the spatial level of SMSAs in the US - concludes that: 'In the U.S., the largest urban areas are the greatest agglomerations of industrial research and development' (pp.323). On the basis of a regression analysis, he even concludes that the relationship between number of industrial R&D *laboratories* (Y) and population (X) 'is quite strong (r=0.97), but is nonlinear, with an increasing number of R&D facilities found in the larger cities' (pp.323). This relationship is depicted in Figure 2.3 below.

Figure 2.3 Population and R&D Laboratories

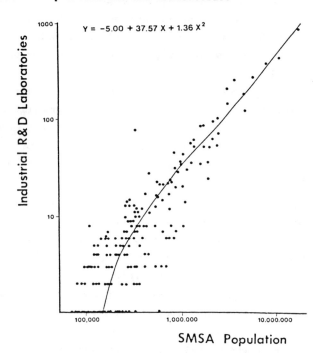

SMSA Population

Source: Malecki 1979a, pp.325

In the *UK*, Howells (1984) disentangles the same kind of spatial orientation of *total R&D employment* - again including the public sector - towards the South East. However, on an urban-rural basis, i.e.,

> a classification scheme based upon four categories of counties (conurbation, more urbanized, less urbanized and rural ... it is the less-urbanized counties which dominate the spatial pattern of research and development service employment rather than the conurbations (Howells 1984, pp.17).

Concerning the *changes* in the spatial dispersion of total R&D employment in the period 1971-1976, a clear deconcentration tendency - especially at the urban-rural scale - is revealed by the data.

2. Studies which analyze the spatial dispersion of R&D activities in large *high tech* (multi-locational) firms (cf., Malecki 1979b, 1980, Oakey 1984, Howells 1984, Gibbs and Thwaites 1985). The 'bias' towards these large 'high tech' firms can be explained by two factors. Firstly, these firms perform the *greater part of private R&D activities* in Western countries nowadays. Secondly, the issue of *data availability* should be mentioned. R&D data relating to *small* (independent) firms are generally scarce and would require more in-depth analyses (cf., Oakey 1984 and Kleinknecht 1987b).

The *spatial* dispersion of R&D activities in these large firms are largely dependent on the functional *corporate organization* of these firms (cf., Malecki 1980, Howells 1984 and Gibbs and Thwaites 1985). In this respect, it is useful to distinguish between basic, applied and development oriented research. In general, 'complex' R&D functions (i.e., basic and applied research) are expected to be - both functionally and spatially - more centralized than development oriented R&D activities. These latter types of R&D facilities will often be linked to - and support - *production oriented* facilities.

Consequently the spatial dispersion of head offices and production plants is considered to be important in determining the locations of more fundamental and development oriented research facilities, respectively. Regarding fundamental R&D facilities, however, Malecki (1980) *also* accentuates location in *innovation centres* - i.e., 'university cities with some federal scientific activity and some manufacturing' (Malecki 1980, pp.231). Especially in a functionally *decentralized* organization structure of R&D, such locations often become attractive as 'A decentralized structure has the greatest number of locational possibilities for R and D, whereas centralized R&D is only infrequently found in locations other than the corporate headquarters' (Malecki 1980, pp.225).

On the basis of interviews and a postal survey in large 'high tech' firms - i.e., pharmaceuticals, metal working machine tools, scientific and industrial instruments and radio and electronic components - the analysis of Gibbs and Thwaites (1985) confirms the above sketched functional and spatial organization of R&D activities. In these firms

> centralised research is (indeed) concentrated within the core regions, particularly the South East, for ease of contact with corporate headquarters and both government and MoD[14] research institutions. Development work is much more widely distributed and is predominantly associated with production plant (Gibbs and Thwaites 1985. pp.13).

The empirical evidence presented by Malecki (1980) also refers to large multi-locational 'high tech' firms - i.e., instruments, aerospace and electrical and electronics products - known to perform significant R&D efforts.[15] His analysis also confirms the spatially biased pattern of R&D laboratories within these firms towards the larger urban areas and 'innovation centres' (like the Boston area or California locations).

As stated above, a methodological problem inherent in the spatial analysis of *R&D activities* in *multi-locational* firms is that the location of 'generation' of innovations (for example, a new product developed by a central R&D facility) may differ from the location of 'implementation' (i.e., where the new product will be produced). This applies especially regarding the 'fruits' of the centralized (fundamental) R&D facilities. The spatial effects of *intra-firm diffusion* processes may result in a relatively high degree of innovative activities - i.e., adoption in 'branch plants' - in peripheral regions (cf., Thwaites 1982).

Malecki and Varaiya (1986) however, argue that these latter diffusion processes may be rather limited as - at least in first instance - new products will often be located near the R&D facilities which generated them. This would be caused by the fact that in the experimental phase such products will often have to modified rather drastically and, consequently, intensive communication (i.e., spatial proximity) between R&D and production facility is a necessity.

This hypothesis is confirmed in the empirical analysis of Gibbs and Thwaites (1985) who found that

> The research did not, however, support the proposition that new products are developed in research units or plants in the core and then transferred to the regions. Most innovations are put into production where they were originally developed (Gibbs and Thwaites, pp.15).

On the basis of their interviews and establishment surveys they conclude that especially location *close to R&D* and not headquarters facilities is important.

So a general empirical finding of the spatial dispersion of R&D facilities - in multi-locational 'high tech' firms - appears to suggest a spatially biased pattern of R&D facilities towards the larger metropolitan areas. This applies especially to the more fundamental (basic and applied) R&D facilities. On the other hand, production oriented (i.e., development) R&D facilities are often linked to the production sites and, consequently, display a more decentralized spatial pattern. However, data limitations and confidentiality issues often preclude an in-depth analysis of the spatial dispersion of all (types of) R&D activities.[16]

3. Studies on the spatial dispersion of *R&D activities* in *small and medium sized* 'high tech' firms.[17] Although these firms may also play an important role in the generation of new technologies and industries empirical research relating to (the spatial dimension of) innovation *inputs* in these firms is very scarce (cf., Kok et al 1985, Kleinknecht 1987b). Given the lack of small firm R&D data in general, the application of costly and time-consuming micro-level inquiries appears inevitable. An example of this type of analysis can be found in Oakey (1984) - who also analyzes the 'output aspect' of innovative activities (see below).

Oakey concentrates on small[18] independent 'high tech' (i.e., instruments and electronics) firms in three selected areas -the San Francisco Bay Area (including Silicon Valley), the South East of England and the development region of Scotland. These regions were selected in order 'to study the most diverse regional economic environments with hypothesized qualities ranging from the potentially least conducive to the most conducive for innovation' (Oakey 1984, pp.239), respectively. Concerning the regional incidence of internal R&D the following pattern came to the fore:

Table 2.2   The Incidence of Internal Research and Development by Region

Research and development effort	Scotland No.	%	South East No.	%	Bay Area No	%
Internal R&D	42	77.8	51	85.0	55	91.7
No R&D	12	22.2	9	15.0	5	8.3
Total	54	100.0	60	100.0	60	100.0

Chi square = 4.31, p=0.116, N=174

Source: Oakey 1984, pp.241

The rank order of these regions - regarding the incidence of internal R&D - is in accordance with a priori expectations.[19] However, it is also noteworthy that firms located in more centrally located 'high quality' economic environments are *not sole masters* concerning innovative efforts (as approximated by the regional incidence of R&D). In this respect, one of the main conclusions Oakey reaches is that: 'Scotland has not readily accepted the role of a typical development region in decline' (Oakey 1984, pp.249).

The finding that *also* in *peripheral regions small* independent firms are engaged in *generating* innovations and do not only *adopt* - with a considerable time-lag - new technologies is an important break with 'the initial assumption, that innovative and research activity is concentrated near the centre, or top, of the urban and regional hierarchy' (Howells 1984, pp.14).[20]

2.5.2.3   Innovation Output Studies

In contrast to the above-mentioned types of studies, innovation *output* studies (also) consider the output aspect of innovative activities carried out at the individual firm level (see the left part of equation (2) in subsection 2.2.4). A few methodological problems surround this type of analysis, however.

The *first* more general problem relates to the *measurement* of innovativeness - i.e., the *output* aspect of innovative activities - of individual firms (see also Hansen 1986 and Ewers 1986). In this respect, innovativeness can be considered as a complex multi-dimensional phenomenon which is difficult to define and measure in an exact and precise way. Consequently, several innovation output *indicators* are often used (for example, the number of product innovations generated in a certain period of time, or process innovations, number of patents granted) to approximate this phenomenon. As will be shown later on, these results may differ according to the indicator used.

A *second* problem relates to the *causal analysis* of approximated spatial variations in the innovativeness of firms. As stated in section 2.2.1, such variations may be due to a constellation of 'structural' and 'production milieu' effects. However, empirical validation of the impact of *both components* simultaneously is generally not provided. This may, inter alia, be caused by 'the number of cases which is necessary in order to allow a combined quantitative analysis of size and location characteristics' (Ewers 1986, pp.169).

As a consequence of these methodological problems, Ewers (1986) perceives a 'relative lack of innovation studies allowing a combined analysis of aspatial *and* spatial characteristics on the plant level' (Ewers 1986, pp.169). Some examples of innovation output studies will be presented below. As stated above, several different approximations of '*innovativeness*' are being used.

Oakey et al (1980), for example, combine the Science Policy Research Unit Innovation Data Bank - concerning 1,200 important post-war innovations - and the Queen's Award Data - which awards 'outstanding achievements in British industrial firms by acknowledging either increased exports or new technological innovations' (Oakey et al 1980, pp.237). Both data sets (partly)[21] refer to *significant innovations* in British industry. This study might also be classified in the lower left part of our innovation typology as some of the selected innovations were generated in establishments belonging to the 'old' or 'traditional' sectors. However, the greater part of the selected establishments indeed belong to the 'high tech' sectors such as electrical engineering, chemical engineering, instrument engineering and aerospace.

As stated above, spatial variations in the innovativeness of firms or establishments may be due to both a 'production milieu' and 'structural' impact.[22] This distinction is also (implicit) in the analysis of Oakey et al (1980). From their analysis of the 'structural' component, several intra-establishment variables (e.g., sector, plant size, organization structure) appeared to be relevant. Consequently, the determination of an additional 'production milieu' effect - on spatial variations in the generation of significant innovations[23] - is hampered by the fact that they encounter the second methodological 'size' problem discussed above:

> How far the regional variations may be generally attributed to the differing industrial structures of each region is difficult to ascertain precisely given the limited number of cases in the study (Oakey et al 1980, pp.243).

In this context Thwaites (1982), has analyzed spatial variations in the *generation* of (especially) *product innovations* - i.e., at the establishment level - in the selected 'high tech' sectors mentioned before.[24] As regards *group plants*, spatial variations in the introduction of product innovations appeared to be only marginal in nature. As stated in our discussion of multi-locational firms, this might be (partly) due to intra-firm diffusion processes. On the other hand, 'independent enterprises located in the Development Areas lagged behind those located in other areas' (Thwaites 1982, pp.375).

A statistical estimation of *both* the 'production milieu' and 'structural' impact on innovative performance however, is not provided. Concerning the latter component, especially the South East appears to be in a favourable position as 'large variations in on site R&D activity were observed within the independent enterprise sector where those located in the Development Areas lagged well behind those located in the South East' (Thwaites 1982, pp.376).

Oakey (1984), in analyzing the incidence of product innovations in the ('high tech') data set mentioned above, arrives at the same kind of 'stalemate' position. Although his data set reveals a spatially varying pattern of the introduction of product innovations (see Table 2.3 below), this might largely be due to the 'structural' (intra-firm) variable of 'on-site R&D' (see Table 2.2 above). Table 2.4 below clearly demonstrates this correlation between the *'structural'* (intra-firm) variable of *'on-site R&D'* and *incidence of product innovations*.

Table 2.3   Incidence of Product Innovations in the Five Year
            Period Prior to Survey by Region

Product innovation	Scotland No.	Scotland %	South East No.	South East %	Bay Area No.	Bay Area %
Innovation	34	63	47	78	51	85
No innovation	20	37	13	22	9	15
Total	54	100	60	100	60	100

Chi square = 7.84,   p = 0.019, n = 174

Source: Oakey 1984, pp.239

Table 2.4   Presence of Internal R&D by Incidence of Product Innovation

Research and development effort	Product innovation No.	Product innovation %	No innovation No.	No innovation %
Internal R&D	123	93.2	25	59.5
No R&D	9	6.8	17	40.5
Total	132	100.0	42	100.0

Chi square = 28.4, p = 0.0001, N = 174

Source: Oakey 1984, pp.241

'Patent studies' can also be considered as a variant of spatial innovation output studies. These studies approximate regional variations in the output of innovations by means of patent data (cf., Ewers et al 1979, Kok et al 1985). In as far as 'patent studies' are not limited to specific new 'high tech' sectors,[25] these studies can both be located in the upper and lower left part of Figure 2.1 in section 2.2.

Ewers et al (1979), for example, demonstrated that in the mechanical-engineering sector in West-Germany, patent rates were four times higher in non-assisted areas compared to assisted ones. Concerning patent data in general however,

there remains that doubt which arises from the deficiencies of patents as indicators for innovation performance: not all patents lead to marketable innovations, and by no means all innovations are based on patented innovations (Ewers and Wettman 1980, pp.167).

Especially this latter point becomes increasingly problematic nowadays as - given the increased speed of technological changes and the time-lag between the application for and granting of a patent-patenting becomes a less effective means for appropriating the 'fruits' of innovative activities (cf., Nelson 1986, Camagni and Rabellotti 1986). On the basis of their analysis of 'high tech' firms in the Milan area Camagni and Rabellotti (1986), for example, conclude: 'As regards patents, we have realized a narrow attitude to patent in electronics industry: the speed of the innovation process makes the patents redundant' (pp.20).

After having discussed spatial innovation studies which can be classified in the *upper left* part of our innovation typology, we shall subsequently discuss the lower left part of this typology.

2.5.3 Innovative Activities in Existing ('Traditional') Sectors

Spatial innovation studies which (mainly) analyze regional variations in innovative activities performed in more 'traditional' or non-'high-tech' sectors are relatively scarce.[26] As stated above, these studies appear to be strongly biased towards (large) 'high tech' firms. The argument that these firms can be considered as prime 'generators' of technical advances is often used to defend such 'bias'.

In the *'sector studies'* discussed above (see subsection 2.5.2.1) it is especially the spatial dispersion of 'high tech' sectors which is being analyzed. In these studies, a spatial bias towards non-'high-tech' sectors is *not* considered to be conducive to the relative innovative performance of a region.

Also the *'innovation input'* studies concentrate especially on (large) 'high tech' firms. This point is also stressed by Alderman and Thwaites (1987). However, they also argue that especially peripheral areas - given their 'bias' towards more 'traditional' sectors - may be largely dependent on innovative activities performed in these sectors. For this reason, they analyze the spatial dispersion of R&D activities in a more 'traditional' sector, i.e., machine tools. Their empirical results are based on the same survey results as used in Thwaites (1982) and on a follow-up to this survey (undertaken at the end of 1986).

Regarding the spatial variations in the ('structural') variable of 'on site R&D in 1981', the patterns are largely[27] consistent with the above mentioned patterns in the UK - i.e., with East Midlands, East Anglia and the South West leading and Scotland, Wales and the Northern region lagging. Regarding the spatial *changes* in the incidence of *'on site R&D'* among 'survivors' - i.e., establishments identified in the 1981 data set and still existing in 1986 - however, they conclude that as regards full time R&D 'it would seem that there has been a diminishing of the regional differences' (Alderman and Thwaites 1987, pp.9).

Kleinknecht and Mouwen (1985) - in analyzing R&D activities within the manufacturing sector as a whole in the Netherlands - also observe a *spatial deconcentration* tendency of these activities to 'intermediate' regions.

In Chapter 7 of the present study spatial variations in the innovation potential of various types of - not specifically 'high-tech' - manufacturing establishments will be analyzed for the Dutch context.

With the above mentioned reservations in mind - concerning the analysis of Oakey et al (1980) and 'patent studies' - also the *'innovation output'* studies tend to be biased towards (large) 'high tech' firms. Innovation output studies concentrating on the innovative performance of, for example, 'old line' firms are rather scarce. Some examples in this respect are Kok et al (1985) - although they do not in particular concentrate on 'old line' industries (see below) - and Davelaar and Nijkamp (1989d). This analysis will also be considered in Chapter 6 of the present study.

Kok et al (1985), consider innovative activities within small firms - i.e., with 100 employees or less. In this context, they do not in particular concentrate on 'high tech' firms. One of the main purposes of their analysis is the identification of spatial variations in the innovativeness of firms - both manufacturing and services - in the Netherlands. On the basis of sample-data, for example, they compared the degree of innovativeness of firms located in metropolitan areas (i.e., Rotterdam, Utrecht and Arnhem/Nijmegen) with the 'average' innovative performance in the Netherlands as a whole. Although some evidence is provided that firms located in these metropolitan regions exhibit a marginal higher level of innovativeness, this applies especially to 'low quality' types of innovations. At a regional scale of analysis - i.e., considering the 'scores' of the provinces in the Netherlands-especially 'intermediate' provinces like Overijssel and Gelderland appeared to perform rather favourable. Concerning the intra-firm determinants of innovativeness several variables (like sector, firm size, age of the firm) appeared to be relevant. Whether - besides these 'structural' or intra-firm variables - also the 'production milieu' component has any relevance is not being analyzed in this study by Kok et al, however. Like in many studies discussed above, a multi-variate analysis determining both the impact of the 'structural' and the 'production milieu' component on spatial variations in innovativeness is not provided.

In this and the preceding section we have discussed several *types* of spatial innovation studies as distinguished in section 2.2. In the next section, the most important conclusions of the present chapter will be recapitulated.

2.6 CONCLUSION

In this chapter, the spatial dimension of technological activities has been discussed. It has been demonstrated that also in spatial analysis innovative activities are increasingly acknowledged to be of utmost importance for the economic performance of firms and, consequently, regions. Essentially, the *spatial* dimension enters the scene at two levels.

Firstly, spatial variations in the innovative performance or innovativeness of firms may be caused by the fact that regions differ according to types of firms - i.e., on the basis of intra-firm characteristics. In this context, it will be clear that regions which are relatively well endowed with, for example, large 'high tech' and/or advanced service firms will be in a favourable position regarding the number and quality of techno-economic innovations generated. This spatial variation in the *types* of firms located in a region has been labelled the '*structural*' component.

Secondly, location in a favourable 'production milieu' - ceteris paribus the intra-firm characteristics or 'structural' component - is often considered to affect the 'innovativeness' of firms.[28] This (external) spatial impact has been labelled the *'production milieu'* component.

As discussed in this chapter, both 'diffusion' - i.e., the second type of diffusion research discussed above - and 'generation' studies often indicate spatial variations in innovative performance of firms. In this respect, *empirical* evidence concerning the impact of *both* the 'structural' and the 'production milieu' component on these variations is very scarce, however. As regards the statistical determination of both these components, the main methodological problems involved refer to the *measurement* of such concepts like 'innovativeness', 'innovation potential' and 'production milieu', as well as the *number of (micro-level) cases* necessary for such an analysis.

Consequently, a statistical determination of the (external) 'production milieu' impact on innovativeness - i.e., after compensating for the 'structural' (intra-firm) component - is generally not provided. As a consequence, 'evidence' concerning the impact of the 'production milieu' component is still largely 'suggestive' in nature. In this context, Thomas (1987) remarks: 'Current knowledge of the role of the regional factor's influence on entrepreneurial activity is primarily based on case histories and anecdotal evidence and there is a great lack of empirical verification' (pp.25).

Consequently, a *first basic research issue* of the present study will be:

1. *To determine the impact of both the 'structural' and 'production milieu' component on spatial variations in the innovative performance of firms.*

As will be clear, this will imply that the methodological problems identified above will have to be tackled. This research issue is important because the positive (additional) impact of location in a 'favourable' production environment is often implicitly assumed. On the basis of this assumption several attempts have been made to 'approximate' the 'quality' of the 'production milieu' component in several regions or cities and (implicitly) consider this to be a 'proxy' for spatial variations in the 'innovativeness' of individual firms. Given the lack of empirical validation of this 'production milieu' impact however, the empirical validation of this hypothesis appears to be a useful research endeavour. Chapter 6 of the present study will particularly deal with this research issue.

As regards the *first* level at which the spatial dimension enters the technological scene - i.e., regional variations in the 'structural' component - the following remarks are in order. In innovation *diffusion* analysis the main emphasis is placed upon a description (or estimation) and explanation of the space-time trajectory of specific new technologies or 'entrepreneurial' innovations. Spatial variations in *economic performance* - resulting from innovative activities - are expected to arise from spatial variations in *time-lags* of *adoption* of such innovations. In this analysis the role of technological change is largely limited to an *exogenous* generator of such innovations. Consequently, the role of *continuous* technological change - i.e., of *endogenously generated* innovations as an effective means of competition between firms and regions - is only barely considered in diffusion analysis. The ignorance of the 'creative' and the *endogenous* mechanisms inherent to innovative activities within firms and regions is one of the basic imperfections of this type of analysis.

The notion that firms are capable of *endogenously generating* further innovations is implicit in those spatial innovation studies which have been classified under the heading of '*generation*' studies. These studies (implicitly) acknowledge the role of endogenously generated innovations as an effective means of (spatial) competition.

On the other hand however, these studies largely ignore - contrary to the innovation *diffusion* studies - the spatial *dynamic* aspects of technological change. As stated before, these 'generation' studies are especially biased towards the spatial dispersion of innovative activities performed in selected - mostly (large) 'high tech' - firms and establishments *at a certain moment in time*. Consequently, these studies concentrate on the question where - at a certain moment of time - the 'bulk' of innovative activities regarding certain selected firms or sectors is located in space. In this context, one of the themes so basic in diffusion analysis - i.e., whether the locus of innovative activities (adoption) regarding new technologies shifts over time[29] - is generally disregarded.

In the foregoing chapter, the *dynamic* and *'creative'* aspects of (further) innovative activities (along technological 'trajectories') have been stressed. In that chapter the importance of the *'creative diffusion'* process - in generating new life cycles of industries and technologies - has also been pointed out.

From these observations, it appears to be a fruitful research endeavour to integrate both spatial innovation approaches - which can be considered complementary rather than competitive - with the findings of the preceding chapter in a *spatially oriented 'creative diffusion' framework.*

In the next chapters, we will try to develop such a framework. In this respect, our main purpose will be to provide a theoretical-conceptual framework concerning the 'generation' studies discussed above, although also 'diffusion' elements will be incorporated. As will be clear from the foregoing discussion this appears to be a useful research endeavour as these 'generation' studies are empirically oriented and a lack of a theoretical framework can be observed (see also the next chapter). Consequently, our *second basic research isssue* will be:

2. *To develop (and test empirically) a theoretical-conceptual framework concerning new life cycles of industries and technologies in a spatial context.*

This research issue both involves both the *dynamic* and *'creative'* aspects inherent in spatial diffusion and spatial 'generation' studies, respectively. Consequently, as regards this second research issue the following three research questions will be considered in this framework:

2a. Where do the life cycles of new industries and technologies - or new types of firms linked to new technological 'trajectories' - start in space?

2b. Does the locus of (further) innovative activities - within *such industries* - shift in the course of time?

2c. Is there any regional specialization tendency concerning *types of* innovative activities performed in the life cycles of such industries and technologies?

In the next two chapters, we will start with the further investigation of the three questions related to our second basic issue identified above.

Notes to Chapter 2:

1. This is a regional classification of the Netherlands in 40 rather homogenous zonal regions.

2. Or, stated in terms of Chapter 1: firms or sectors linked to more recent technological 'trajectories'.

3. This is the cumulative proportion of adopters at a certain moment in time.

4. Consequently, this more recent version of the urban incubation hypothesis does not refer to the generation of new firms in general (for details on the urban incubator hypothesis the reader is referred to Davelaar and Nijkamp 1987a).

5. This applies in the case of a decentralized decision making structure (cf., Brown 1981).

6. In this respect, we will ignore the possibility of a 'random' diffusion pattern.

7. In this context, Brown (1981) however, also indicates the importance of the *strategy* pursued by the new innovative establishment.

8. Or, stated in terms of Chapter 1 of this study, the spatial consequences of the 'creative diffusion' process are largely ignored in this type of diffusion analysis.

9. For example, the relevance of specific new pieces of technology will differ according to the type of firm. Consequently, the regional dispersion of different types of firms (i.e., the 'structural component') will influence spatial variations in adoption.

10. As stated in Chapter 1 of this study these further innovative activities relate to both the product (service) characteristics and the related process technologies.

11. At least as far as employment is concerned.

12. Data concerning the spatial dispersion of employment of establishments are often - at least concerning the larger establishments - *integrally available*. See also Chapter 9 of the present study.

13. This is caused by the fact that R&D data are difficult to collect.

14. Ministry of Defence.

15. Malecki includes those firms for which R&D data are available concerning 1977 and the location of R&D facilities are known.

16. In the Netherlands, for example, the Central Bureau of Statistics collects only R&D data concerning the larger firms. Because of confidentiality even these limited data are not avaible at a (disaggregate) spatial level.

17. Also from a *spatial* point of view, innovative activities in these firms may be especially important. The economic effects of innovations generated in such small firms will often favour the region in which these firms are located (given the fact that many small firms will be single-plant).

18. In this case, firms employing less than 200 employees were included.

19. In this respect, the Bay Area appears to be leading while Scotland lags behind.

20. Inherent to, for example, diffusion research as discussed in the foregoing section.

21. When considering the Queen's Award Data.

22. As stated before, the 'structural' and 'production milieu' component may - at least in the long run - be interdependent. Given the cross-section approach applied in this type of research, however, this interdependency is generally ignored.

23. Especially East Anglia, South-West and South-East England performed relatively favourable.

24. See the analysis of Gibbs and Thwaites (1985) mentioned above. In this respect, the term 'high tech' applies especially to the scientific and industrial instruments and radio and electronic sector.

25. Patents are not registered according to sector. This 'translation' from patents to sectors often appears to be hazardous (cf., Kok et al 1985).

26. Recall again the fact that the innovations identified in Oakey et al (1980) and 'patent studies' may (partly) refer to the more 'traditional' sectors.

27. With the exception of the South East which appeared to have an 'average' performance.

28. Or, stated otherwise, the 'productivity' of the intra-firm innovation inputs such as R&D efforts.

29. This 'ignorance' is also reflected in the bias towards relatively new ('high tech') sectors at the beginning of their life cycle and the lack of spatial innovation studies referring to innovative activities performed in later phases of the life cycles of industries and technologies.

3 New technological 'trajectories' and spatial dynamics

3.1 INTRODUCTION

In the previous chapter, several types of spatial innovation studies have been discussed. In this respect, we have especially concentrated on the conceptualization of technological change within these studies. As pointed out in that chapter, in *diffusion* analysis main emphasis is placed upon the *spatial dispersion* of exogenously generated ('entrepreneurial') innovations *over time*. As can be concluded from our discussion in Chapter 1, this approach bears several similarities to more traditional approaches in economics in which technological change is treated as 'manna from heaven'. Malecki and Varaiya (1986), for example, state that in this type of analysis technological change is treated as a kind of 'black box' phenomenon.

Concerning the *spatial dimension*:

> Large urban areas are expected, ceteris paribus, to have higher rates of innovation, more rapid adoption of innovations, and higher proportions of skilled workers than smaller places, but technological change itself is not endogenous (Malecki and Varaiya 1986, pp.633).

The '*generation*' studies on the other hand - in line with more recent trends in economic analysis - do not consider technological change as a largely exogenous phenomenon. On the contrary, in these studies it is (implicitly) acknowledged that the *rate and direction* of technological change can be *manipulated* by means of, for example, R&D efforts. In this respect, the *generation of innovations* itself is considered to be an important means of gaining a competitive edge over one's competitors.

As discussed in the preceding chapter however, these 'generation' studies are *empirically oriented* and *theoretical analysis is scarce.* This statement is also affirmed by Malecki and Varaiya (1986) in their review of spatial innovation analyses: 'empirical research on technology and regional structure and change has progressed more boldly and broadly than theoretical approaches (pp.642). Concerning these 'generation' studies Thomas (1987) and Nijkamp (1987) also underscore the *need of a theoretical framework* as research 'has been intuitive and inductive and lacks the guidance of a sound theoretical framework' (Thomas 1987, pp.25). In a similar vein, Nijkamp (1987) remarks: 'The locational analysis of new technology firms is still underdeveloped: a clear theoretical framework is lacking' (pp.277).

Given these considerations, in this and the next chapter a *first attempt* to design such a framework will be made. The development of this framework was the purpose of the second basic research issue already identified in the previous chapter. In the present chapter, a general outline of such a theoretical-conceptual framework will be sketched. In the next chapter, this framework will be illustrated by means of a theoretical (simulation) model.

As stated before, our conceptual framework will especially apply to the 'generation' studies discussed before, although also 'diffusion' elements will be incorporated. In this context, various recent approaches regarding the role and characteristics of technological change - as discussed in Chapter 1 - will be considered. Based on a macroscopic view, the spatio-temporal dynamics related to the emergence of new 'technology systems' and the resultant 'take-off' of new technological 'trajectories' will receive main focus (cf., Freeman et al 1982, see also Chapter 1). In this respect, the relationships between *new life cycles of industries and technologies* will be *linked* to the *spatial dimension.* Consequently, in our framework the *role of technological change* in shaping and re-arranging *spatial structures* will be considered.

Our framework bears some resemblance - but also several dissimilarities - to the well-known spatial product-life cycle approach (cf., Vernon 1966, Hirsch 1967). Therefore, in the next section we will start with a brief discussion of this concept. Basically, we will concentrate on the role of technological change within this framework. Next, in section 3.3 we will present some of the main criticisms- especially regarding the incorporation of technological change - which can be raised against this concept.

This discussion will set the stage for our conceptual-theoretical framework to be discussed in subsequent sections. To this purpose, first in section 3.4 the general points of departure concerning our framework - based on the foregoing chapters - will be outlined. Next, several stages of the spatial life cycles of (new) industries and technologies- and the role of technological change - will be considered in section 3.5. Various testable hypotheses will be deduced in section 3.6. Empirical validation of these hypotheses will follow in the second part of this study. Also in this last section the relationship of our framework with other spatial innovation research efforts - as discussed in the previous chapter - will be considered. In section 3.7, the main conclusions concerning this chapter will be summarized.

3.2 THE PRODUCT LIFE CYCLE APPROACH IN A SPATIAL CONTEXT

In this section, our main attention will focus on a discussion of the *spatial* product life cycle concept. Underlying this concept is the S-shaped growth curve of a new product (or service) characterized by the *slow growth introduction* phase, a *fast growth phase*, followed by a *saturation* and/or *decline* phase as the product matures (cf., van Duijn 1983, de Jong 1985). In *spatial* analysis, this concept is often used in 'explaining' spatial production and/or employment trends (see, for example, Krumme and Hayter 1975, Norton and Rees 1979, Hekman 1980 and Barkley 1988). Consequently, in this section we will present a brief discussion of the spatial product life cycle approach. In the next section, we will discuss several *criticisms* which can (especially) be raised against the conceptualization of technological change within this approach.

Originally, the spatial product life cycle concept has been developed by Vernon (1966) who tried to explain *international investment* and *trade* patterns. In his analysis the life cycle of a new product - or an innovation - can be distinguished in *three phases* each of which will be characterized by *specific locational patterns*:

First stage:

The first stage refers to the initial (production) location of a new product or an innovation. In this respect, Vernon (1966) first of all rejects 'the powerful simplifying notion that knowledge is a universal free good, and introduce it as an independent variable in the decision to trade or to invest' (pp.192). In this context, especially the US - and to a lesser extent other advanced countries - are assumed to possess favourable *market opportunities* for (first) introducing new products.[1] Given the fact that knowledge (concerning these market opportunities) is considered to be *non-ubiquitous*, this implies that 'United States producers are likely to be the first to spy an opportunity for high-income or labor-saving new products' (Vernon 1966, pp.194).

Although *production* of the innovation need not necessarily take place near the (US) market, Vernon identified three forces (see below) which might attract initial production to central (i.e., US) regions. These forces 'are stronger than relative factor-cost and transport considerations' and relate to 'communication and external economies' (pp.194). In the initial phases of the introduction of a new product, the nature of the design, the market and the necessary inputs are still unstandardized. Consequently, initial production is marked by the following *needs* and *characteristics* (see Vernon 1966):

1. The want of *flexibility* concerning the *inputs* used. This need follows from the fact that in the initial phases, the inputs used will often have to be changed quite frequently.

2. In the initial phases, prices concerning the inputs used (see above) are less important given the rather *low price elasticity of demand* prevailing in these phases. Consequently, 'small cost differences count less in the calculations of the entrepreneur than they are likely to count later on' (Vernon 1966, pp.195).

3. In initial phases, main emphasis is placed on *introducing* the new product itself to the market. This implies that:'The need for swift and effective communication on the part of the producer with

customers, suppliers, and even competitors is especially high at this stage' (Vernon 1966, pp.195).

Given these observations *initial production* is expected to be attracted to *central (US) regions*. Regarding the first and third 'need' mentioned above, central (metropolitan) regions are generally considered to be in a favourable position. At the same time, the higher input (labour) prices prevailing in these regions (see also below) count less in the initial phases of production.[2]

Second stage:

The second stage refers to the *location* of (the production of) the *maturing product*. As the demand and production for a new product increases, the new product and its production process will become more *standardized* and 'a growing acceptance of certain general standards seems to be typical' (Vernon 1966, pp.196). This implies that the centripetal forces identified above become less important. Besides and above this, at this time demand for the new product may also expand in the 'other advanced countries' which display a delayed adoption pattern as compared to the more central (US) regions.

In this phase also production *prices* become increasingly *important* in determining the competitive performance of firms. In this context, Vernon especially considered *spatial variations in labour costs* as the critical variable determining production cost differences between US and 'other advanced countries'. Consequently, in this phase firms located in the central (US) region may locate part of their production facilities- e.g., by raising 'branch plants' - in other advanced (non-US) countries. So spatial *differences in production costs* - in spite of transportation costs - may cause production for third countries or even US markets to be serviced from production facilities raised in the 'other advanced countries'.

Third stage:

The third stage concerns the *standardized product*. In this phase, access to market information becomes less critical and, consequently, location in *less developed countries* comes within reach. This tendency will be reinforced by the fact that as production and markets are highly standardized they 'tend to have a well-articulated easily accessible international market and to sell largely on the basis of price' (Vernon 1966, pp.203). Concerning the *production costs*, the *low cost of labour* in less-developed countries may (again) exert a positive impact upon international investments in these countries. Therefore part of the production activities will be relocated to these areas.

The predicted locational patterns of the spatial product-life cycle concept discussed above, are depicted in Figure 3.1 below.

Figure 3.1   The Spatial Product Life Cycle

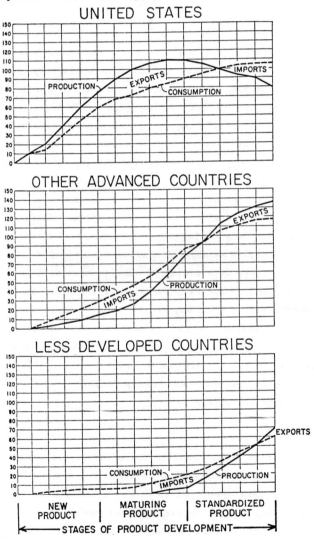

Source: Vernon (1966), pp.199

Although Vernon especially specified the concept for the *international* context, he also mentioned several examples - e.g., concerning the spatial shift of textile and electronics production within the US - indicating that the same kind of mechanisms might also be at work at the *national scale*. In spatial analysis especially this *latter variant* has become popular nowadays (cf., Norton and Rees 1979, Norton 1979, Taylor 1987, Barkley 1988, Karlsson 1988). In this interpretation, the US, 'other advanced countries' and less-developed countries can be 'substituted' by *metropolitan*, *intermediate* and *peripheral* regions within a country, respectively. In these later versions, the product life cycle approach is often not restricted to multi-national firms but it is also used to describe more aggregate (sectoral) patterns of production and employment (see Norton and Rees 1979 and Barkley 1988, for example).

3.3 TECHNOLOGICAL CHANGE WITHIN THE SPATIAL PRODUCT LIFE CYCLE APPROACH; A CRITIQUE

In this section some shortcomings of the product life cycle concept - as applied to the spatial dimension - will be discussed. In this respect, we will mainly concentrate on a discussion of criticisms which refer to the conceptualization of *technological change* within this framework. These criticisms will set the *stage* for *our* (extension of this) *approach* to be discussed in the following sections.

The *first* and main criticism against this concept refers to the incorporation of technological changes regarding the new product. In this respect, it can be stated that 'the treatment of invention and innovation within the model is wholly unrealistic. It is implicit within the product-cycle model that inventions are introduced to the market in their final form' (Taylor 1987, pp.83). As an innovation is (implicitly) assumed to be introduced in its 'final form', a 'Schumpeterian swarming process' in which other (new) firms enter the scene - and are able to capture part of the market by means of (continuously) improving the new products and/or related production processes - is largely disregarded.

In Chapter 1 of this study however, it has been pointed out that the diffusion of the production of new types of products or services is *seldom a carbon-copy process* of imitation, but is frequently characterized by further innovative activities (cf., Rosenberg 1976, Freeman et al 1982). These further improvements apply to both the *product* (characteristics) and the related *process technologies*.[3] These improvements are at the heart of the 'creative diffusion' process discussed in Chapter 1 (see also section 3.4).

Although Vernon (1966) mentioned the possibility of *product differentiation* - also in the 'mature phase' - in more recent interpretations this aspect is often disregarded (cf., Taylor 1987). Also Vernon (1966) however, ignored the spatial competitive effects which might result from such a 'creative diffusion' process.

Also the role of (further) technological changes regarding the related *production process technologies* is only barely considered. However, also concerning the production techniques of new products (and services) further improvements - i.e., *process innovations* - will often be generated. This type of technological progress has especially been stressed in more traditional economic analysis (see Chapter 1).
Regarding capital investments, for example, this often implies that *capital efficiency* increases over time. In this respect, *'early'* producers of a new product will often bear such 'learning economies' and invest in *less efficient 'vintages'* compared to *'late'* producers.

This type of technological change may be especially relevant in a *spatial* context as 'early' and 'late' producers are expected - according to the spatial product life cycle concept - to display a *specific locational pattern*. In the spatial product life cycle approach the spatial implications of *this type* of technological change is (also) generally disregarded, however. In this respect, 'The inertia of capital is too easily forgotten and the assumption of perfect capital mobility can be too glibby applied' (Taylor 1987, pp.81).

The spatial product life cycle concept is especially concerned with specific new products which will first be produced by (existing multi-national) firms in central regions. In later phases production will shift to other regions. Consequently, the role of *fundamental technological (and socio-institutional) innovations* - i.e., new 'technology system' - in generating new life cycles of industries and technologies is not captured by this concept. In this respect, the (long-run) impact of these fundamental (technological) transformations on structural changes within the spatial economy is disregarded.

Central or metropolitan regions are considered to be the *centres of (all types of) innovative activities.* This conclusion follows from the fact that these regions are expected to be both *first adopters* and *producers* of new products or services. The *competitive power* of these regions depends on their ability *to attract innovations* on the basis of communication and external economies (see also below).

Over time however, these economies become less important and part of the production will be decentralized. This shift in production is fostered by *spatial variations in production costs.* These costs become increasingly relevant in later phases of the life cycle of a new product. Especially spatial variations in *labour costs* are considered important in this context. As stated above, the role of further technological changes - concerning new products as well as the related production process technologies - in this deconcentration process is largely ignored.

Consequently, in the spatial product life cycle approach the role of *technological change* is rather *exogenous.* In the initial phase the production of a new product is attracted to core regions mainly because of *communication and external economies* to be gained here. During later phases production will shift from these regions especially because of spatial variations in *input prices* which, by then, become increasingly relevant. Technological change itself however, has little bearing on the *spatial shift* of production and employment.

In this context, the spatial product life cycle concept has severe difficulties in coping with the empirical observed fact (see the previous chapter) that also firms located in *non-metropolitan* areas are engaged in *generating* innovations. This point is also (partly) reflected in the 'explanations' provided concerning the urban-rural shift of manufacturing employment in all EEC-countries (cf., Keeble et al 1983). Ewers (1986), for example, proposes as a possible explanation[4] that 'innovativeness of small firms[5] should be independent of the location or even greater in locations outside the agglomerations' (pp.168). However, such an 'explanation' does not fit the theoretical approaches - e.g., diffusion models and the spatial product life cycle concept - in spatial innovation analysis. So, in this context, Keeble et al (1983) derive three possible explanations from the recent literature as regards this spatial shift of manufacturing employment:

a. *The production cost explanation* which 'focusses attention on the impact of presumed substantially higher operating costs for manufacturing industry in urban centres' (pp.406). Especially spatial variations in wage and salary costs are presumed to be important in this context.

b. *Constrained location theory* which 'focusses attention almost exclusively on the impact on manufacturing employment in cities of factory supply constraints' (pp.406). This approach especially stresses the physical congestion effects in urban areas.

c. *The capital restructuring approach.* 'The key aspects of the approach focus on capital accumulation and centralisation in large multiplant corporations, and a shift of production by such firms to areas characterized by exploitable, unskilled and cheap labour' (pp.407).

Keeble et al (1983) conclude that these explanations 'clearly overlap at various points, most notably with regard to the significance of labour costs in the production cost/filter-down theory and capital restructuring viewpoints' (pp.407). Consequently, both the first and third 'explanation' mentioned above are largely in accordance with the spatial product life cycle approach discussed above. As can be concluded from these 'explanations', the fact that this shift has something to do with *innovation* is *generally ignored*, however.

The role of (regional) *demand* patterns is only limited in the spatial product life cycle approach. In general, Say's Law is implicitly adopted as 'the market is always assumed to be able to absorb all that is produced, even in the fact of increasing competition' (Taylor 1987, pp.81). Consequently, the interdependencies between local innovative activities and regional demand patterns remain largely out of scope.

In Chapter 1 however, it has been discussed how *market saturation* concerning the (new) technological 'trajectories' generated after World War II has been an important cause of the recession in the seventies and early eighties. As a result of this market saturation economic recovery largely depends on the generation of new types of products and services filling in - and opening up - new markets. Consequently, innovative activities and demand patterns are interrelated.

As to the *spatial* dimension, this phenomenon of market saturation may again be highly relevant. Both *adoption and production* - according to the spatial product life cycle approach - are hypothesized to *start from central or metropolitan* regions onwards. This implies - assuming an equal speed (though not timing) of adoption within the regions - that producers in central regions will also be the *first* ones noticing (signs of) *market saturation* as regards their own region. Consequently, also *market opportunities* will display a *spatio-temporal pattern*. The spatial impacts of such demand patterns - and their interaction with local innovative activities - are largely disregarded in the spatial product life cycle concept, however.

To summarize the above arguments, the main criticisms regarding the incorporation of technological change within the spatial product life cycle concept refer to:

<u>1</u>. The relationship between *fundamental* technological (and socio-institutional) innovations - or new 'technology systems' - in generating new technological 'trajectories' and resultant *structural changes* in the spatial economy. Or, stated otherwise, usually the *spatio-temporal* dimension of new life cycles of industries and technologies are ignored in this type of analysis.

<u>2</u>. The treatment of *technological change* as regards (the characteristics of) new types of products or services. Innovations are (implicitly) assumed to be introduced in their 'final form'. In this context, the spatial implications of a 'creative diffusion' process - in which (new) 'Schumpeterian' firms enter the scene and conquer part of the new markets by further improving these products or services - are largely ignored.

<u>3</u>. The incorporation of technological changes concerning *production process technologies*. In this context, (continuous) improvements with respect to production techniques - i.e., process innovations - and their embodiment in (fixed) capital goods are generally left out

of consideration. This also applies to the resultant (spatial) inertia.

<u>4</u>. The interdependencies between (spatial variations in) demand and (local) innovative activities are only barely considered.

As stated above, each of these 'shortcomings' may have clear *spatial implications*. The above-mentioned criticisms will provide the *building blocks* of our own theoretical framework to be discussed in the next sections. It will be especially regarding the conceptualization of *technological change* that our framework will deviate from the concept discussed above.

3.4 ASPECTS REGARDING ECONOMIC AND TECHNOLOGICAL CHANGE

In this section, our main points of departure regarding the *characteristics* of *technological change*[6] and its economic impacts - and vice versa - will be presented. In the *next sections*, these points will be linked to the spatial dimension in a conceptual-theoretical framework. Our points of departure will largely be based on our synopsis presented in Chapter 1. Consequently, they will only be briefly (re)considered here:

<u>1</u>. *First of all, the (discontinuous) emergence of new 'technology systems' - i.e., a constellation of basic and socio-institutional innovations - will be considered as one of the important vehicles of structural changes in the (spatial) economy.*

<u>2</u>. *The (spatial) economic effects of such innovations will be realized through 'swarming processes' of (new) 'Schumpeterian' firms trying to generate new types of products and services to be derived from these innovations.*

<u>3</u>. *Technological change concerning (the generation of) these new types of products and services often proceeds along underlying technological 'trajectories'.*

In this respect, the 'swarming processes' discussed above can be considered as 'creative diffusion' processes resulting in further

(secondary) innovations. In general, the attraction of (new)
'Schumpeterian' firms does not imply a 'carbon-copy' process of
imitation of the original innovators. However, this attraction
frequently involves the generation - and further improvement of the
new products or services[7] (i.e., product innovations) and/or the related
process technologies (i.e., process innovations). In the sequel, these
'paths' of further innovative activities will also be denoted as
'normal' technological progress (cf., Nelson and Winter 1977).

4. *The attraction of many (new) 'Schumpeterian' firms and the
resultant further innovative activities will result in new life
cycles of industries and technologies.*[8]

5. *As innovative activities are often (highly) interdependent - i.e.,
proceed along specific technological 'trajectories' - this will result
in technological 'learning economies'.*

These 'learning economies' refer to both the (characteristics of
the) *new products and services* generated and the related *'production
techniques'*. These 'learning economies' are at the heart of the
'creative diffusion' process in which *(new) 'Schumpeterian' firms* will
try to become competitive by means of *further improving* the newly
generated products (and services) and/or the related process
technologies.

6. *At the individual firm level, technological change is often
characterized by several 'inertia' phenomena and 'irreversibilities'.*

Firstly, firms do not perform innovative efforts in a vacuum, but
rather these activities are *linked to prevailing production activities*
(cf., Nelson and Winter 1982). In this respect, by far the greater part
of the R&D efforts of firms are oriented towards a further improvement
of products or services and/or the related production processes which
are *in line* with their *current activities*.

For example, as regards 'high tech' firms located in the Milan
area Camagni and Rabellotti (1986) conclude on the basis of their
inquiry that

> In all firms, the innovation activity was found to be represented
> by a continuous flow of small, incremental improvements of the
> product lines. Within this innovative flow, however, it is possible
> to identify the 'core products' and the main technological

trajectories (pp.19-20).

This implies, for example, that the R&D investments of firms are not fully flexible (and costless) adjustable.

Secondly, the related *production process technologies* will often be embodied in *fixed capital goods*. As will be clear, also these fixed capital outlays will result in (a certain degree of) inertia at the individual firm level.

7. *Both the notion of continuous technological change - along (specific) 'trajectories' - and the (technological) rigidities at the individual firm level imply that in many cases the extraordinary profits derived from innovative activities will only be temporary.*

In this respect, the technological 'learning economies' achieved so far will often 'spill-over' (in later phases) to other (new) firms. These 'spill-over' effects are at the heart of the Schumpeterian *'creative diffusion'* process discussed in Chapter 1.

Especially the initial phases of the life cycles of industries and technologies will be characterized by a high level of new firm formation - e.g., by means of 'spin-offs' (cf., Rothwell and Zegveld 1985; see also before). In these phases other (new) firms - attracted by the exceptional profit possibilities - will often be able to capture part of the expanding markets by means of *reaping the fruits* of the (technological) *learning economies* achieved so far. In this respect, they will try to gain a competitive edge by means of generating *further* improvements as regards the manifest most successful new products or services. In this context, Malecki (1979a) remarks:

> the greatest risk lies with the innovator; adoption may be more successful, even though later, for firms which imitate but concentrate their R&D on product improvement (pp.322).

8. *Concerning the types of innovative activities performed the notion of the innovation life cycle concept is important (see Chapter 1).*

According to this concept, in the course of the 'creative diffusion' process the main emphasis of further innovative activities will *shift from product to process innovations*. As stated in Chapter 1, in the inital phases of the 'swarming processes' many *product innovations* will be generated. In the course of time however, it will become increasingly difficult and costly to generate - or further improve - new products and services. Consequently, in the course of the 'creative diffusion' process innovative efforts aiming at improving the related *production techniques* - i.e., *process innovations* - will become increasingly important as an effective means of (technological) competition.

9. *Demand factors are relevant in two respects, viz., as market pull factors and bottlenecks.*

Firstly, although technology-push factors may dominate in the initial phases of the 'creative diffusion' process, in the course of time *demand pull* factors will become increasingly important in determining the direction of further innovative activities (cf., Freeman et al 1982, Morphet 1987).

Secondly, in the course of time *limits* regarding the possibilities for further *market expansion* - e.g., by means of technological improvements - will be approached. In this context, it becomes increasingly difficult and costly - i.e., because of the decreasing productivity of R&D efforts - to raise the marginal utility of new or improved products or services. Consequently, both *technological* and *demand* factors - i.e., market saturation - will set *limits* concerning the expansion of new industries and technologies.

In the next section, the issues discussed above will be linked to the spatial dimension.

3.5 TECHNOLOGICAL CHANGE IN A SPATIAL CONTEXT

3.5.1 'Creative Diffusion' in Space

In this section, it will be discussed how life cycles of new industries and technologies are expected to *evolve over space* and result in *spatial structural changes*. However, it should be stressed at the outset that we do *not* intend to deny the role of other 'driving forces' - identified in the literature - in generating spatial dynamics. *Our main purposes relate to (also) stressing the role of (fundamental and further) innovative activities in generating spatial structural changes.* So, basically, our approach will deviate from prior spatially oriented research efforts in *two respects*.

First of all, radical changes in technology and the socio-institutional framework - i.e., the emergence of new 'technology systems' - are expected to lead to (discontinuous) structural changes in the spatial economy. In this context, the 'bunching' of the 'take off' of new technological 'trajectories' - following the emergence of a new 'technology system' - and the 'swarming processes' of (new) 'Schumpeterian' firms attracted to these 'trajectories', will lead to new spatial life cycles of industries and technologies.

Consequently, in a long-run perspective, the economic structure of regions is not constant, but is *being re-shaped by such (new) life cycles*. In spatial innovation analysis, the (long-run) interrelations between changing spatial structures and (the 'bunching' of) new technological 'trajectories' have largely been neglected.

Secondly, the spatial structural changes resulting from the 'creative diffusion' processes along such new technological 'trajectories' will be considered. This aspect refers to the often *continuous nature* of technological change regarding new types of products and services. From our discussion in the foregoing chapter it can be concluded that also this aspect of technological change is only barely considered in spatial analysis. For example, most of the *spatial diffusion models* analyze the diffusion of 'innovations' which are hypothesized to be both '*static*'[9] in time and space. Also in *spatial product life cycle analysis* the spatial economic impacts of a 'creative diffusion' process are generally disregarded.

It will be the purpose of this section to make a *first step* towards the integration of both these notions of technological change in spatial analysis. In the next chapter, this framework will be further illustrated by means of a theoretical (simulation) model.

In the following pages, we will develop a '*dynamic incubation*' *framework* in which *both notions of technological change* identified above act as major '*driving forces*' of *spatial dynamics*. The general outlines regarding this framework will be illustrated below.

As to the spatial dimension, it is *not* our purpose to '*explain*' the mere *emergence* of - and *timing* of the 'swarming processes' along - new technological 'trajectories'. The *discontinuous* nature of these 'swarming' processes is generally agreed upon, however.

Consequently, our spatial analysis will mainly concentrate on the *spatial implications* of both (the 'bunching' of) the 'take-off' of *new technological 'trajectories'* and *'creative diffusion'* along these 'trajectories'. In this respect, the following three phases will roughly be distinguished:

- The incubation phase
- The catching-up or competition phase
- The stagnation phase

The (ideal-typical) locational characteristics of these phases will be discussed in the next subsections.

3.5.2 The Incubation Phase

This phase refers to the *initial stages* of the *'swarming processes'* along new technological 'trajectories'. As stated before, these new 'trajectories' will be related to (a constellation of) underlying basic and socio-institutional innovations, i.e., a new 'technology system'. The entrance of many (new) Schumpeterian firms in these phases - generating further innovations - will result in new life cycles of industries and technologies.

As to the *spatial* dimension, the study of locational advantages of central (metropolitan) areas - regarding the initial adoption and generation of new economic activities in general - is rather well established in spatial analysis. It would go too far - and beyond the purposes of our analysis (see subsection 3.1) - to deny the existence and relevance of these locational advantages.

Consequently, in this stage of the analysis we will adhere to the theoretical spatial approaches - e.g., spatial diffusion analysis, the product life cycle approach and the more recent version of the urban incubator hypothesis[10] - discussed before. In our analysis, this implies that we expect that in the initial stages the 'swarming processes' of (new) 'Schumpeterian' firms attracted to the new 'trajectories' will especially be biased towards the larger central (metropolitan) complexes.[11] Especially in these initial stages such firms will be able to benefit from 'advantages' to be gained in these regions.

Several of these locational advantages have been (extensively) discussed in spatial analyses and also in prior sections of this book. In short, these advantages refer to the *four clusters* of *'production milieu'* variables identified in subsection 2.2.3 of Chapter 2:

1. *Agglomeration economies* caused by the spatial clustering of (dis)similar firms. In the initial phases of the 'swarming' process many new (types of) products and services - i.e., product innovations - will be generated. Given the 'immature' nature of these products and services, *flexibility* concerning the inputs is often a prerequisite (cf., Vernon 1966, see section 3.2). Because of these agglomeration economies, especially metropolitan or central areas are able to provide this flexibility.

2. As regards the *demography and population structure*, the availability of *skilled labour* will be highly relevant. Especially in the initial phases of the new 'trajectories' skilled labour is crucial (cf., Freeman et al 1982). Also this aspect will largely favour metropolitan or central areas as highly skilled labour is often attracted to such areas.

3. The availability of *specialized information* and *intensive communication* patterns (cf., Hoover and Vernon 1959, Pred 1977 and Lambooy 1984) - i.e., producer-user relations between 'consumers' and 'suppliers' of the newly generated products and services - can be mentioned as a third factor favouring metropolitan or central areas.

4. The 'swarming' processes along the newly established 'trajectories' often involves *new types of demands* regarding *social overhead capital* (cf., Brown 1981, Moss 1985; compare for example the current need of telecommunication networks). In general, such new types of social overhead capital will *first* be developed in *metropolitan or central areas*. The density of economic activities will favour the *remunerativeness* of such investments in these areas.

Consequently, in our theoretical framework the *spatial structural changes* - resulting from the 'swarming' processes along newly established 'trajectories' - are in first instance expected to be especially effectuated in the larger metropolitan complexes.

Metropolitan or central areas are expected to be important *incubation areas* of new 'Schumpeterian' industries *in the initial phases of the 'swarming' processes along newly established technological 'trajectories'.*[12] As regards *types of* innovative activities, the initial bias (of 'normal' technological progress along new 'trajectories') towards product innovations implies that - as regards *these industries-* many *product innovations* will be generated in these areas in the initial phases.

3.5.3 The Catching-up or Competition Phase

The catching-up or competition phase refers to the adoption - and the concomitant 'swarming' processes - of the new 'trajectories' in intermediate and peripheral areas respectively.[13]

As the 'swarming' processes along the new 'trajectories' proceed, possibilities for *significantly improving* the new (types of) products and services generated in earlier phases of the 'swarming' processes will (somewhat) *decrease.*[14] 'Normal' technological progress - i.e., the further innovative activities along the new 'trajectories' - will *increasingly be oriented* towards (minor) product or service improvements and - especially and increasingly more towards - *process innovations.*

Consequently, in the later stages the further innovative activities will become less turbulent. Concomitantly, the importance of the *locational 'pull' factors* discussed above - as regards metropolitan areas - will also *decrease.* In this respect, for example, the necessary social overhead capital may become increasingly available in non-metropolitan areas. As a corollary, the ability of (entrepreneurs located in) intermediate and peripheral areas, respectively, to take part in the new 'trajectories' will *increase.*[15]

This *spatial deconcentration* tendency will also be fostered by the fact that *markets in metropolitan or central areas* will - as a consequence of first adopting the new 'trajectories' - first approach (signs of) *market saturation* concerning several of the new (types of) products and services generated. As stated before, this market saturation may result from two intertwined forces. Firstly, 'consumers' in these areas - whether households or firms - often take the lead in the adoption of innovations (see, for example, the diffusion models and theories discussed in the preceding chapter). Secondly, the (new) 'Schumpeterian' firms initially attracted to these areas will often concentrate their production activities on this growing *local* 'consumer' market (cf., Vernon 1966).

From these arguments it follows that in later phases of the 'swarming' process possibilities for (further) *market expansion* will especially exist in *non-metropolitan or non-central* - i.e., intermediate and peripheral - *areas*.

Consequently, as the 'swarming' process proceeds, both *demand* and *supply factors* will become *less advantageous to metropolitan or central areas*. Several of the newly established technological 'trajectories' will become less tightened to metropolitan or central areas and will also be taken up in intermediate and peripheral areas, respectively.

This *spatial reorientation* of the production activities related to the new 'trajectories' may be effectuated, through *relocation*, the raising of *'branch plants'* and *endogenous new firm formation*.[16] Given the inertia resulting, inter alia, from spatially fixed capital investments, local labour force and local market areas, *interregional relocation* will often be too costly. Besides and above this, the rather favourable profit possibilities generally prevailing in the initial phases of the 'take-off' of new 'trajectories' will often render this option less necessary. *Empirical evidence* also points at a rather minor effect of the interregional relocation component in changing spatial structures *in general* (cf., Bluestone and Harrison 1982, Wever 1984). The importance of the *'branch plant'* component will, inter alia, be determined by the degree in which - as regards certain new 'trajectories' - oligopolistic market structures will be established.[17]

Concerning those new 'trajectories' which are *not* permanently and completely dominated by a limited number of firms the regional indigenous component - i.e., local (new) firms attracted to the new 'trajectories' - will be (especially) important in this spatial deconcentration tendency. In the following pages we will especially concentrate on the *spatial implications* of this latter type of spatial 'reorientation' (i.e., 'trajectories).[18]

In this case, the 'swarming' processes will become increasingly biased towards intermediate and peripheral areas, respectively. As discussed before however, these processes do not proceed as a 'carbon-copy' process of imitation, but frequently involve the generation - or further improvement - of new products or services (i.e., product innovations) and/or related process technologies (i.e., process innovations).

In this context, these late 'adopter' (new) firms in non-metropolitan or non-central areas will be able to *reap the 'fruits'* of the technological 'learning economies' achieved so far. They will benefit from economies in *two respects. Firstly*, concerning the new products and services generated in the earlier phases of the 'swarming' processes - in (especially) metropolitan areas - they will concentrate on the *further improvement* of those products and services which have proven *to be most successful. Secondly*, technological 'learning economies' also apply to the *production processes* of the newly generated products and services. As stated in section 3.4 these 'learning economies' will often result in more efficient production process technologies over time. Consequently, these late 'adopter' firms will also *invest in* and/or *further improve* the most efficient (i.e., recent) production *process technologies.*

Consequently, *in later phases* of the 'swarming' processes - along newly established technological 'trajectories' - the inherent 'Schumpeterian' *dynamics* will shift to *intermediate and peripheral areas*, respectively. Given the 'creative' nature of these 'swarming' processes however, this implies that also the *locus* of *further innovative activities* - as regards these industries and technologies-will *shift* to these areas. This 'technological dominance' in these later phases will (further) enhance the competitive power of these areas. As discussed before (see section 3.4) however, in the course of the 'creative diffusion' process the main *emphasis* of these further innovative activities will *shift from product to process innovations.*

Consequently, although in later phases of the 'swarming' processes non-metropolitan (or non-central) areas may become dominant concerning the innovative activities performed, as regards the *types of* innovations generated (or adopted), this 'dominance' will especially - and increasingly more - apply to *process innovations*.

Consequently, in our framework *non-metropolitan or non-central areas* will *also* play their (specific) *innovative 'roles'*, in *later phases* of the *'swarming' process* along new technological 'trajectories'. In our framework, the increased *competitive performance* of these areas in *later phases* of the 'swarming' process will *not* (only) *depend on a-technological* factors - such as favourable input (labour) prices, less unionization, less congestion and so on (cf., Thompson 1968). *On the contrary, it will also depend on (further) local innovative activities along (newly established) technological 'trajectories'*.

In Figure 3.2 below, the *ideal-typical* pattern of the innovative performance of *metropolitan or central areas* - i.e., in the course of the 'swarming' processes along new technological 'trajectories' - as compared to the 'average' innovative performance of other regions has been sketched for both product and process innovations.[19]

Figure 3.2   Innovative Performance of Metropolitan Regions concerning
New Technological 'Trajectories' in the Course of Time

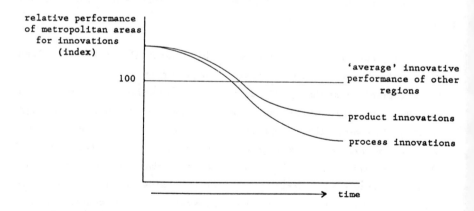

3.5.4 The Stagnation Phase

In the course of time, several of the new (types of) products and services generated along the newly established 'trajectories' will be confronted with *market saturation* in *all regions*. In this context, both *supply* and *demand* aspects will set boundaries concerning the further expansion of the related new industries (see section 3.4). Consequently, the rate of new 'Schumpeterian' firm formation will level off in all regions concerned and *price competition* will become increasingly important as an effective means of competition.[20]

In this context, as regards the technological dimension, 'late adopter' firms in intermediate and (especially) peripheral areas will, ceteris paribus, be in a favourable position as compared to 'early adopter' firms in metropolitan areas. Because of investing in - and concentrating their innovative efforts on generating - more recent (efficient) production techniques, these 'late adopter' firms (on average) will possess the *most efficient process technologies*.

Consequently, possibilities for *low production costs* in these regions may not (only) be *caused* by lower labour costs. On the contrary, it will (also) result from the spatial implications of the 'creative diffusion' process as regards production process technologies.[21] This aspect of technological change will further enhance the competitive performance of 'late adopter' firms in intermediate and peripheral areas in these later stages. Therefore, the markets of firms located in metropolitan areas will increasingly be 'captured' by these late 'imitators'.[22]

Consequently, although the new industries may become stagnant as a whole, this will especially apply to firms located in *metropolitan or central areas*. Just like these areas were the *first to adopt* the new 'trajectories' and the resultant new life cycles of industries, they will also be the *first to see them shrinking*. The *inertia* embodied in a *fixed (technology) structure* affects the competitive power of these areas in *later phases* of the 'swarming' process so that a process of decline concerning these industries will also start here.[23]

To conclude this section, the spatial *hierarchical ordering* of new economic activities assumed in our theoretical framework - i.e., the spatial ordering of the 'swarming' processes along newly established technological 'trajectories' - is not uncommon in spatial analysis. The hierarchical component in spatial diffusion analysis and the product life cycle approach, for example, generate the same spatial patterns of (producer) adoption.

Concerning the *technological dimension* however, these concepts largely ignore the *spatial implications* of the '*creative diffusion*' process. In these notions the product and/or services diffused among producers remain constant and the role of continuous technological change - as a driving force of spatial dynamics - is largely ignored. However, in case 'normal' technological change proceeds along specific 'trajectories' this will result in further innovative activities. These activities are at the heart of the Schumpeterian 'creative diffusion' process and cannot be ignored in spatial innovation analysis. This issue will be further illustrated in the next chapter. First of all however, in the next section some testable hypotheses will be deduced from our framework sketched above.

3.6 SPATIAL IMPLICATIONS

3.6.1. Testable Hypotheses

In this subsection some testable *hypotheses* will be *deduced* from our dynamic incubation framework sketched in the previous section. Clearly, data limitations - concerning the spatial life cycles of new industries, the spatio-temporal dimension of innovative efforts within such industries and types of innovations generated - preclude an in-depth and complete *empirical testing* of this conceptual-theoretical framework.

However, it is possible to deduce several hypotheses from this framework which can be tested *empirically*. The selection of such hypotheses will be the purpose of this section. The hypotheses which will be deduced in this section, relate directly to the research questions concerning the *second basic research issue* identified in the previous chapter. In the second part of this study we will turn to the empirical testing of these hypotheses. Consequently, these investigations will provide modest and partial tests concerning the empirical validity of our theoretical framework.

These testable hypotheses - related to our theoretical framework - are the following:

<u>1</u> First of all, in the foregoing chapters we have discussed how the emergence of new 'technology systems' - i.e., a constellation of basic and socio-institutional innovations - will lead to structural (sectoral) changes in the economic system.
In this respect, we expect metropolitan or central areas to take the lead in such structural transformation processes. Or, stated otherwise, such areas are expected to take the lead in both the growth and decline of (new) life cycles of industries and technologies.

The empirical testing of this hypothesis for the Netherlands will - inter alia - be considered in Chapter 5 in which - by way of empirical illustration - the (timing of the) *structural economic changes* in the *Amsterdam* area will be compared with structural changes in the Netherlands as a whole.

<u>2</u> *Secondly, our framework suggests a 'hierarchical' spatial diffusion pattern of new 'Schumpeterian' industries linked to more recent 'technology systems' or technological 'trajectories'.*

This hypothesis will be tested in Chapter 9 in which the *spatial diffusion* of some *highly dynamic* and *knowledge-intensive* producer service subclasses - linked to the more recent information 'technology system' (cf., Freeman 1987a, 1987b) - will be analyzed for the Dutch context.

<u>3</u> Thirdly, both metropolitan and non-metropolitan areas are expected to play their *own specific* (see the fourth hypothesis below) *innovative roles* in *different phases* of the *life cycles* of (new) industries and technologies.
This implies that in later phases of the life cycles of industries and technologies non-metropolitan or non-central regions may become dominant concerning the innovative activities performed.

In this respect, we expect metropolitan or central areas to perform relatively favourable in initial phases - as regards employment, production and innovations - concerning firms (sectors) linked to more recent 'trajectories'. On the other hand, these regions are expected to perform relatively poorly - because of the *spatial patterns* of 'creative diffusion' along these 'trajectories' - as regards sectors linked to more *former* 'trajectories'. Concerning peripheral areas, for example, the opposite conclusions are expected to hold.

This hypothesis can be summarized in Figure 3.3 below which relates the evolution of spatial structures to *both* the spatial effects of new 'trajectories' and 'creative diffusion' along such 'trajectories'.

Figure 3.3 'Creative Diffusion' along Technological 'Trajectories' in
Space

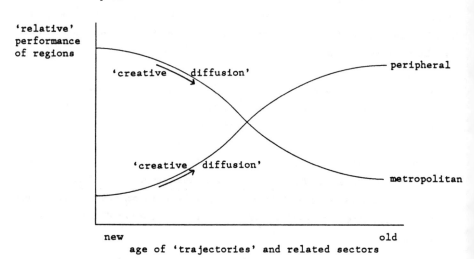

In this figure 'relative' performance implies the 'score' of a region - regarding different sectors (for example, regional sectoral share of employment or innovative activities) - as compared to some *average* or 'normal' *level* (for example, regional share in total employment). For sake of simplicity, we have sketched the ideal-typical performance of a metropolitan area vis-a-vis a peripheral area for sectors linked to 'trajectories' of different periods.

In Chapter 2 of this study it has been discussed that empirical analysis has especially concentrated on the *locus of innovative activities* within (large)[24] 'high tech' firms. In this respect, the notions 'high tech' and firms 'linked to recent technological trajectories' will be largely *overlapping*. Consequently, the (often) empirically observed spatial concentration of innovative efforts - as regards 'high tech' firms - in *central or metropolitan regions* (see Chapter 2) is in accordance with our own framework.

The question whether similar conclusions also apply to *other types* of (e.g., non-'high tech' manufacturing) firms is far less considered in spatial analysis, however (see Chapter 2). On the basis of our framework it would be expected that innovative activities in such firms will display a (much) more *spatially dispersed* pattern. Consequently, in order to test this hypothesis in Chapters 6 and 7 of this study more general spatial innovation patterns with respect to the manufacturing sector will be considered. In these chapters *spatial variations* in the *innovation potential* of several types of - especially *small and not specifically 'high tech'* - manufacturing firms will be analyzed for the Dutch context.

4   *Fourthly, although in later phases of the life cycles of industries and technologies non-metropolitan or non-central regions may become dominant concerning the innovative activities performed, as to the types of innovations generated this 'dominance' will especially apply to the generation of process innovations.* This hypothesis follows logically from the assumption that as 'normal' technological progress - or the 'creative diffusion' process - along new 'trajectories' proceeds, these latter types of innovations will become increasingly important as an effective means of technological competition (cf., the innovation life cycle concept).[25]

This fourth hypothesis will be considered in Chapter 8 of this book. In that chapter, it will be tested whether - for several regional classifications - any regional specialization tendency concerning the *types of innovations generated* - i.e., product vis-a-vis process innovations - can be disentangled in Dutch manufacturing firms.

3.6.2 Relationship with other Spatial Innovation Research Efforts

Before testing the hypotheses deduced above, it should be stated that our framework is embodied in - or supported by - several (types of) spatial innovation research efforts discussed in the previous chapter:

1. The *spatial pattern* of *sectors* linked to technological 'trajectories' of different periods (see Figure 3.3 above) is largely in accordance with the empirical findings concerning the spatial distribution of 'high tech' sectors discussed in Chapter 2.[26] As stated before, 'high tech' can - too a large extent - be considered synonymous with 'sectors linked to more recent technological 'trajectories'. Consequently, the often observed pattern that 'high tech' sectors - as regards *employment* or *number of establishments* - are oriented towards metropolitan or central areas, and poorly represented in peripheral areas, fits our own framework.

2. As stated in the context of the *third hypothesis* above (see subsection 3.6.1), this conclusion also applies to empirical studies pointing at the *metropolitan bias* of *innovative activities* within such 'high tech' firms (see Chapter 2).

3. Freeman et al (1982) describe the (ideal-typical) *sectoral* adoption pattern of (specific) *high quality* and broadly applicable *new technologies* in manufacturing. These innovations are often the focus of analysis in the second type of spatial diffusion analysis described in the foregoing chapter. In general terms, the timing (and rate) of *adoption* of such technologies proceeds from sectors linked to *new* '*trajectories*' (i.e., 'high tech' sectors) to sectors linked to *former* '*trajectories*'.

Given the (expected) *spatial* distribution of these sectors - see again Figure 3.3 above - the sectoral adoption pattern as described by Freeman et al (1982), will *largely favour initial adoption patterns in metropolitan areas*. Consequently, also the (often) empirically observed metropolitan bias - as regards the adoption patterns of such innovations - fits the theoretical framework sketched above.

4. In Chapter 1 and section 3.4. of this study it has been pointed out that the more *significant* (product) innovations will especially be generated in the *early phases* of new technological 'trajectories'. In later phases, emphasis will shift towards further (minor) improvement and process innovations. Given the *expected initial bias* of (firms linked to) *new 'trajectories'* to metropolitan or central areas,[27] this implies that also the *more radical innovations* will especially be generated in these areas. This supposition is confirmed by those 'innovation output studies' - as discussed in Chapter 2 - which trace the *'incubation areas'* of selected *'high-quality'* innovations (see, for example, Oakey et al 1980).

5. In our framework, several *types of regions* are expected to play their own (specific) *innovative role* in different phases of the life cycles of new industries and technologies. Consequently, the *'overall'* *spatial pattern* of innovative activities - i.e., across all sectors ('trajectories') - will display a *complex* pattern. In this context, it is not to be expected that a specific (type of) region will dominate concerning the generation of *all (types of) innovations in all sectors*.

This supposition is, for example, confirmed by the analysis of Kok et al (1985) who analyzed the spatial distribution of innovative activities in several (i.e., not specifically 'high tech') sectors in the Netherlands. In this analysis, no clear spatial concentration of innovations could be disentangled.

6. The *spatial deconcentration* (of employment or establishments) in manufacturing sectors in *general* (cf., Keeble et al 1983, Wever 1985) as well as in several *'high tech'* sectors (cf., Markusen et al 1986, Barkley 1988) is also in accordance with our theoretical framework. As to the *causes* of this spatial deconcentration however, our framework also *calls attention* for the *role of technological change itself*. Or, stated otherwise, the *spatial implications* of the *'creative diffusion'* process along technological 'trajectories'.

3.7 CONCLUSION

In this chapter, the second basic research issue identified in the previous chapter has been considered by developing a conceptual-theoretical framework concerning the role of *technological change* in generating *structural changes* in the *spatial economy*. In this framework the role of technological change is *twofold*.

Firstly, the emergence of new 'technology systems' - i.e., a constellation of basic and socio-institutional innovations - will lead to the emergence and 'take-off' of *new technological 'trajectories'* concerning new (types of) products and services opened up by these innovations. Because of the high (initial) profit possibilities many (new) 'Schumpeterian' firms will be attracted to these 'trajectories' trying to generate (further) innovations. This will result in new life cycles of industries and technologies.

As to the *spatial dimension*, especially *metropolitan or central* areas are expected to be the initial *incubation areas of such new industries*. This also, and accordingly, applies to the (great many) new products and services - i.e., *product innovations* - which will be generated in these initial phases.

Secondly, the 'swarming' processes along such new 'trajectories' are *creative* in nature resulting in technological 'learning economies' and concomitant further innovative activities. In this context, (new) 'Schumpeterian' firms entering the new 'bandwagons' will *reap* the fruits of these *'learning economies'* achieved so far. Consequently, they will try to gain a competitive edge by means of further moving along these underlying 'trajectories', i.e., by means of (further) improving the most 'successful' new types of products or services and/or their related production process technologies.

Because of both supply and demand considerations, intermediate and (especially) peripheral areas will generally display a delayed pattern of 'adoption' as regards these new 'trajectories'. From a *technological point of view* however, this implies that *especially these regions* will benefit from the *'creative diffusion'* processes in *later phases* of the life cycles of new industries and technologies.

Consequently, as the 'swarming' processes of (new)

'Schumpeterian' firms attracted to the new 'trajectories' proceed, the locus of (further) innovative activities will shift to non-metropolitan or non-central areas. This will enhance the competitive power of ('late adopter' firms in) these latter regions as opposed to metropolitan areas in later phases. Although in later phases of the life cycles of industries and technologies non-metropolitan regions may become dominant concerning the innovative activities performed, as to the types of innovations generated this 'dominance' will especially apply to the generation of process innovations.

Consequently, the emergence and 'take-off' of new technological 'trajectories' - and the resultant new life cycles of industries and technologies - are at the heart of the initial leading position of metropolitan or central areas as regards the adoption of these industries and the concomitant structural changes. In a similar way, the spatial patterns of the 'creative diffusion' process along these 'trajectories' are at the basis of the initial decline of these industries in metropolitan or central areas. In terms of Chapter 2, this supposition implies that the 'structural' component of regions-concerning specific sectors ('trajectories') - is not constant but changes itself in the course of the life cycles of new industries and technologies.

Because of the embodied continuous 'creative diffusion' process along (new) technological 'trajectories', our framework implies a more 'optimistic' point of view for non-metropolitan areas than would be suggested by the product life cycle or spatial diffusion theories and models. Implicitly, these concepts suggest that technological progress (concerning new products or services) has come 'at rest' before shifting to non-metropolitan areas. This ignores the creative and innovative role (and the resultant competitive power) these regions might still perform in later phases of the 'swarming' processes, however.

In our framework, the 'creative diffusion' process is critical to the spatial shifting of both (mutually reinforcing processes) of production and innovation and not only of production in the sense of the spatial concepts mentioned above. Consequently, our framework is clearly in contrast with the usual theoretical considerations in which central (metropolitan) regions are often considered to be the *general centres* of (all types of) innovations.

In our framework, several types of regions are expected to perform their own *specific innovative role* during the *'life cycles'* of (new) industries and technologies. The evolution of regional economic performance related to these 'trajectories' is assumed to *mirror* the *timing of this 'creative role'*.

Consequently, the spatial shift of production and employment in later phases of the 'swarming' process is not generated *in spite of* innovative activities - uniquely favouring metropolitan or central areas - but also *because of* specific *further local innovative activities* in non-metropolitan areas. In our conceptualization, continuous *technological change itself* - along (new) 'trajectories'- is an important *(endogenous) driving force of spatial dynamics*. In the next chapter, our theoretical framework will be further illustrated by means of a theoretical (simulation) model.

Notes to Chapter 3:

1. Vernon (1966) especially referred to high-income or labour-saving innovations. In later versions, this 'constraint' is often not strictly applied anymore, however (cf., Taylor 1987).

2. See the second issue mentioned above.

3. This will result, in more efficient capital 'vintages' in the course of time (see also the next chapter).

4. Ewers rejects this possible explanation on the basis of data referring to *spatial variations in adoption of specific new (high quality) technologies* (like CNC machine tools, microprocessors) within the FRG (i.e., the second type of diffusion research discussed in the previous chapter).

5. These firms appeared to perform better - i.e., in terms of employment - than large firms within the manufacturing sector (see Ewers 1986).

6. As stated in chapter 1, these *technological* characteristics may especially refer ⌐⌐ the manufacturing sector.

7. Concerning new products Taylor (1987) remarks: 'the initial introduction of a product to the market is followed by progressive modification and improvement and invention itself is most frequently incremental rather than revolutionary. The product that begins a cycle may bear little resemblance to the product that ends it, if an end can be discerned at all' (p.83).

8. This 'bunching' of 'swarming' processes will result in new types of (manufacturing or service) industries. This should not be interpreted as (immediately) resulting in new SIC-sectors, however. In general, the statistical registration of such new industries will be delayed. Also and accordingly, these new (types of) industries will often be accommodated - at least in the initial phases - in *existing* SIC-categories.

9. This applies to both their product (or service) characteristics and the related production techniques. In this context - and also in the following pages - 'diffusion models' refer to the *first* type of diffusion studies - i.e., the diffusion of 'entrepreneurial innovations' - discussed in the foregoing chapter unless stated otherwise.

10. For details on the urban incubator hypothesis the reader is referred to Davelaar and Nijkamp (1987a).

11. This hypothesis refers to metropolitan or central areas *in general* and should not be interpreted as referring to each *individual* metropolitan area, separately. Being such an area is no guarantee for attracting a large share of such new activities in the initial phases of the 'swarming' process. This attraction will depend on the initial locational requirements of such industries as well as the 'score' of a specific metropolitan area on these requirements.

12. This implies a restriction of the incubation hypothesis in its original context (cf., Hoover and Vernon 1959; for details on this hypothesis see Davelaar and Nijkamp 1987a).

13. In this respect, we expect the diffusion of these new 'trajectories' and the related 'swarming' processes to proceed largely along *hierarchical* lines. So the 'adoption' pattern in peripheral areas will (again) be delayed as compared to intermediate areas.

14. Concerning several of the new types of products and services generated some *'dominant design'* (cf., Sahal 1981) will emerge.

15. Also the importance of the 'non-technological' factors discussed in the context of the 'urban-rural' shift of manufacturing employment may become increasingly important in these later phases.

16. This latter component also includes existing (local) firms changing their main activities by shifting towards new 'trajectories'.

17. Concerning those 'trajectories' in which the technological 'learning economies' can be appropriated - i.e., in a prolonged way- within the individual firm level, this will result in more 'oligopolistic' market structures. However, it would be irrealistic- especially in the initial phases of the 'swarming' processes - to assume that all new 'trajectories' opened up by the often broadly applicable basic innovations will be completely and permanently dominated by a few (large) firms.

18. The spatial organization of (innovative activities of) multi-locational firms is rather well-documented in the literature (see Pred 1977, Malecki 1979a, 1980, Thwaites 1982; see also Chapter 2 of this study).

19. In this figure, the performance of metropolitan areas - for both product and process innovations - has been scaled against the average performance of other regions in each time period.

20. This may result in more oligopolistic market structures in as far as large scale production will be fostered by scale economies.

21. As a matter of fact, on a *national* scale of analysis the validity of the argument referring to regional variations in labour costs can be seriously questioned.

22. This hypothesis is validated by the analysis of Johansson and Larsson (1989) for the Swedish context. In their analysis they demonstrate that sectors in which price competition prevails are especially biased towards non-metropolitan or peripheral areas.

23. This does not imply that all the initial 'adopter' firms in metropolitan areas 'die'. Several of these firms may be able or indeed be 'forced' - given the high bid-rents prevailing in these areas as well as their high-skilled labour force - to 'switch' to more recent 'trajectories' (generating high values added).

24. As stated before, one of the driving forces behind *large* firm formation is the ability to appropriate the fruits of the 'learning economies' along technological 'trajectories' within the individual firm level (in a prolonged way). Consequently, such firms may often be able to 'withdraw' themselves from the effects of the 'creative diffusion' process described above (for a long period of time).

25. This hypothesis will be considered in Chapter 8 of this study.

26. See in particular subsection 2.5.2.1. of Chapter 2.

27. See the second hypothesis deduced in subsection 3.6.1. or Figure 3.3.

4 A theoretical model for a 'dynamic incubation' framework

4.1 INTRODUCTION

In the foregoing chapters the *ongoing* nature of technological change - as regards new types of products (and services) - has been pointed out. In this respect, it has been stressed that the diffusion of an innovation among producers is seldom a *'carbon-copy'* process of imitation. Other (new) 'Schumpeterian' firms will often try to become *competitive* by *further improving* the new product or service (characteristics) and/or its production techniques resulting in a process of *'creative diffusion'*. Regarding these further innovative activities - related to new types of products (and services) - we used the notion of technological *'trajectories'*.

The *'clustering'* of the 'take-off' of such new 'trajectories'- following the emergence of new 'technology systems' - has been pointed out in the previous chapters also. This 'bunching' of new 'trajectories' and the related 'swarming' processes of (new) 'Schumpeterian' firms attracted to - and trying to move further on such new 'trajectories'- will lead to new *life cycles* of *industries and technologies*.

In Chapter 2 of this study, it has been pointed out that in *spatial* theories and models these aspects of technological change- e.g., concerning the spatial implications of the 'creative diffusion' process - are largely ignored. This 'ignorance' is, for example, also reflected in the causal 'explanations' provided concerning the- empirically observed - spatial deconcentration (of employment in) the manufacturing sector in all EEC countries (cf., Keeble et al 1983). In this context, these 'explanations' provided often exclude - as a logical consequence of (implicitly) abstracting from such further innovative activities - the *role of technological change* itself. In this respect, the suggested 'explanations' are often *a-technological* and refer to congestion effects, capital restructuring, pollution regulations, lower wage rates and unionization rates in peripheral areas, and the like (cf., Vernon 1966, Thompson 1968, Keeble et al 1983; see the previous chapter).

As a *first step* towards bridging this 'technology gap' in spatial analysis, we developed our dynamic 'incubation' framework discussed in the foregoing chapter. Consequently, in our framework *both* the spatial structural changes resulting from (the 'bunching' of) *new* technological 'trajectories' *and* the spatial implications of further innovative activities along such 'trajectories' are considered.

In the present chapter, the general outline of this theoretical-conceptual framework will be illustrated by means of an explanatory model for innovative behaviour. The model mainly serves *illustrative purposes*. In this respect, interdependencies between the following three forces - in generating spatial dynamics (e.g., in production) - will be considered:

1. The 'take-off' - and further product-oriented innovative activities (i.e., improvements) along - a new 'trajectory' related to a new type of *product* (or service). Given the *product-oriented* nature of these innovative activities, they will be denoted as *product innovations* in this study. Taking the example of micro computers, such further improvements of the product (characteristics) might relate to reliability, compatibility, memory-space and speed of the tasks performed.

2. Improvements in the mere *production techniques* of the new (type of) product.[1] As stated in the previous chapter, these innovative efforts become increasingly important in later phases of a new technological 'trajectory'. In these phases, the possibilities for further improving the quality of this new type of product decline, whilst the competition

will increasingly be based on (production) *prices*. Consequently, these innovative activities often aim at decreasing production costs. Given the (production) *process-oriented* nature, these improvements will be labelled *process innovations* hereafter.

3. The spatio-temporal dimension of the *demand* - i.e., adoption-concerning a new type of product.

In section 4.3 the three above mentioned components will be integrated in an aggregate spatial innovation model. Before doing so however, in the next section the *general* assumptions concerning this model will be presented. Next, in section 4.4 various results of computer simulations will be presented, while in section 4.5 the effects of changing several parameter values will be discussed. Finally, some general conclusions concerning this chapter will be presented in section 4.6.

4.2 GENERAL ASSUMPTIONS

In the present chapter we will commence with an '*ideal-type*' of spatial development pattern[2] as regards a new technological 'trajectory' related to a new type of product.[3] The *general outlines* concerning our illustrative and theoretical model will be discussed in the present section. More *specific assumptions* will be considered in the next section in which our model will be specified.

4.2.1 General Outlines

1. Although not strictly necessary for our analysis, we will assume as a point of departure the existence of a '*clustering*' of the emergence and 'take-off' of new technological 'trajectories' in time (see also Chapters 1 and 3 of this study). The idea of clustered 'trajectories' is depicted in Figure 4.1 below.

Figure 4.1   New Technological 'Trajectories' in Time

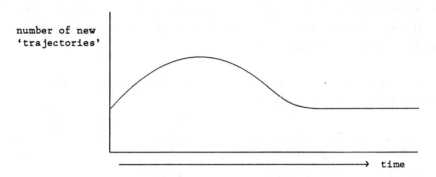

As stated above, in the following pages we will concentrate on an ('ideal-type' of) spatial dynamics concerning *one* specific new 'trajectory'.

2. In order to avoid unnecessary complications in the analysis, we will make a simple trichotomic spatial subdivision into a central (or metropolitan) region (c), an 'intermediate' region (i) and a peripheral region (p). For the ease of presentation - but without loss of generality - we will assume that each area contains an *equal absolute level* of economic activities (e.g., households, firms), but differs in the *spatial density* of these activities. Consequently, the spatial concentration of economic activities for each zone is different and declines from the centre. Thus, for our illustrative purposes, we will assume the following regional classification of a country or region (see Figure 4.2 below)

Figure 4.2   Trichotomic Spatial Subdivision

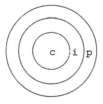

So, if we define D_r as the density of economic activities in region r (r = c, i, p), it is assumed that $D_c > D_i > D_p$. In general, the central region will be metropolitan in character; the intermediate zone will normally show a mixed urban-rural character, while the peripheral zone will be rural in character. For a country like the Netherlands, for example, such a kind of regional classification is quite common (see the next chapters). In this particular case, the centre is more or less synonymous to the Dutch Rimcity (or Randstad); the intermediate zone represents approximately a ring around the Rimcity, whilst the peripheral zone consists more or less of the (rural) regions in the northern, eastern and southern parts of the Netherlands.

3. In our model, the interdependencies between regional production, (types of) local innovative activities and regional demand patterns will be considered. In this respect, it will be assumed that producers flexibly *adjust their production* (capacity) *to the level of demand* accruing to them in a certain period. On the other hand, this level of demand will again be related to *both product and process innovations* of a producer, on the basis of investments in the most recent (efficient) capital 'vintages'.

4. In order to facilitate the analysis, it will be assumed that each region distinguished above possesses *one* entrepreneur willing and able to provide a given new type of product A.[4] As a matter of fact, these entrepreneurs can be considered to be *representative* for similar developments taking place with regard to *several entrepreneurs* - within a specific *type of region* - at *the same time*.

5. In our model the (overall) pattern of *production* in the intermediate - and even more so for the peripheral - area will be delayed as compared to the central region. This is in accordance with the theoretical framework sketched in the foregoing chapter. As pointed out in this framework, central or metropolitan regions are expected to serve as important incubation areas for the 'take-off' of new technological 'trajectories' (i.e., new life cycles of industries and technologies).

In our model, this overall delayed pattern of adoption at the *producer side* in the non-central regions will result from the spatial *hierarchical* development pattern of the *demand*. Consequently, the density of economic activities in the *central* region - and the concomitant intensive interaction and communication patterns - will be hypothesized to lead to 'early' adoption patterns of (potential) 'consumers' in this region.[5] In a similar vein, the low density of economic activities in the peripheral region will cause 'consumers' in this region to be the latest ones in adopting the new type of product.[6]

Given the above mentioned *general* assumptions, in the next section we will incorporate these assumptions in a behavioural model for spatial innovation patterns to be used for simulation experiments on spatio-temporal dynamics related to new technological 'trajectories'.

4.3 THE CONCEPTUAL MODEL

In this section, we will present our model concerning the ('ideal-type' of) spatial dynamics resulting from the emergence of a new type of product and the (inherent) further innovative activities along the underlying technological 'trajectory'. In this respect, we will especially concentrate on *illustrating* the interdependencies between the three forces - identified in the introduction of this chapter - in generating *spatial fluctuations of production over time* (see also Davelaar and Nijkamp, 1989c).

For the sake of simplicity, in the following pages the new type of product will be considered as a *final good* A^7 which is still open to modifications (i.e., improvements). They may relate to both production *process technologies* - i.e., process innovations - and the *product* (characteristics) *itself* - i.e., product innovations.[8]

As a point of departure we will first describe the (aggregate) *regional development of demand* (i.e., adoption) concerning the new type of product A. Although - especially at the beginning - the new type of product itself is still open to several modifications, for the sake of simplicity we will *ignore* the impact of such modifications upon the development of *aggregate* (regional) demand. However, such improvements will *influence* the *allocation* of this *demand* among the various producers of this product (see equation (2)).[9]

Although several formulations might equally well be appropriate in this context, we will apply a Verhulst-type of difference equation in order to describe the development of regional demand concerning the new type of product. Our Verhulst-type equation is rather simple while still generating the - often empirically observed - S-shaped adoption curve (cf., Griliches 1957, Berry 1972, Davies 1979, Metcalfe 1981 and

Blommestein and Nijkamp 1987):

$$DX_{rt} = a_{rt} * DX_{rt-1} * (N - DX_{rt-1}) \qquad\qquad (1)$$

in which:

DX_{rt} = aggregate regional demand for the new type of product A in region r (r=c, i, p) at time t. In this context, regional demand will be set equal to 1 in case $a_{rt} > 0$ for the *first* time (see also Eq. (10)).

N = some absolute maximum level of (potential) demand, which is assumed to be non-region specific; this may be a reasonable assumption in view of our regional classification in which each region contains the same *absolute* level of economic activities (see the foregoing section).

a_{rt} = a measure of the 'willingness to adopt'.

Consequently, it is assumed that in period t-1 → t the effective demand in period t-1 in region r (i.e., those consumers who adopted the new type of product A in this period: DX_{rt-1}) interacts with the unfulfilled (potential) demand in period t (i.e., the non-adopters: N-DX_{rt-1}). This interaction will result - via the 'willingness to adopt' parameter a_{rt} - in a new level of (regional) demand at period t. In our analysis, the parameter a_{rt} is assumed to be region-specific and *time* and *price* dependent (see Eq. (10)).

As May (1976) among others has shown, the conditions for the existence of a (unique) stable equilibrium in equation (1) are: $1 < a_{rt}*N < 3$. Otherwise various types of non-stable and even completely chaotic patterns may emerge. In our simulation experiments in sections 4.4 and 4.5 we will take account of this necessary stability condition by choosing the parameters N and a_{rt} in an appropriate way.

Thus from the above equation for each period one can derive the total demand for the new type of product A in each region. It is assumed that the consumers of this product are - in *allocating their demand* among the entrepreneurs in the different regions - guided by the relative 'attractiveness' of (the producers in) the successive regions (cf., Blommestein and Nijkamp 1987). In this respect, it is assumed that the 'attractiveness' of (the entrepreneur[10] in) a specific region is determined by its *price level* (including general transportation costs) and by further *product* (oriented) *innovations* the region is able to provide at a certain moment of time. Consequently, although the rate of further product improvements - as regards the new type of product A- is not assumed to determine the *overall* level of demand in a region, it determines the *regional allocation* of this demand.[11]

As regards this regional 'attractiveness' we assume a non-linear relationship of the following type (based among others on Allen and Sanglier, 1979):

$$A_{rr°t} = c_r \left[\frac{1 + Prinn_{rt}}{P_{rt} + d_{rr°}} \right]^{\beta} \tag{2}$$

in which:

$A_{rr°t}$ = 'attractiveness' of region r (in period t) to consumers of the new type of product A located in r°.

$Prinn_{rt}$ = measure of the number (and quality) of the (further) product oriented improvements - concerning the new type of product A - generated in region r at time t.

P_{rt} = ('general') *production price*[12] level of (variants of) innovation A in region r at time t.

$d_{rr°}$ = general - i.e., including information costs - *transportation costs* from region r° to region r per unit of innovation A.

β = elasticity of regional 'attractiveness' with respect to a relative (as opposed to prices: see the denominator of equation (2)) 'innovativeness' measure; this parameter is non-region specific and also reflects the degree of unanimity in the response of the consumers (see below).

c_r = a scale parameter

As can be seen from the specification of equation (2), as a first approximation we have assumed that the effect of further improving (the product characteristics of) innovation A - i.e., generating further product oriented innovations - *'favours'* a region only *one period of time*. Consequently, it is (implicitly) assumed that the other entrepreneurs are able to imitate such improvements with a one-period delay. Of course, more complex patterns of imitation might also be incorporated. For example, this could be done by means of a distributed imitation lag function in which further product oriented innovations favour a producer for *several time periods* (at a decreasing rate). However, for sake of simplicity we will ignore this latter option. For the same reason, we also assume that the *production prices* of the (improved) variants of innovation A do *not* differ *within* a region. As we will see later on however, these ('general') production prices may differ *between* the regions depending on the efficiency of production process technologies applied.

According to Allen and Sanglier (1979) the parameter β is a constant which

> measures the degree of unanimity in the response of the population to the relative attractiveness of the centers. It will depend on the similarity of the choice criteria and on the amount of information concerning the merits of the various centers. For $\beta \to \infty$, we will have a 100 percent orientation of demand towards the most attractive center. For a more realistic value, however, we will have some overlap of the spatial demand cones and somewhat indistinct frontiers between the hinterlands (Allen and Sanglier 1979, pp.261).

Consequently, in our model the *competiveness* of (the producers in the different) regions is dependent on both their (further) *product oriented innovations* generated and their ('general') *production prices*. These prices are again related to production *process-oriented innovative activities* (see below).

In case the producers are always able and willing to adjust their production (capacity) to the demand accruing to them,[13] *total production* of (the entrepreneur located in) region r is equal to the *summation* of the (total) *demand attracted from each region*. As stated above, in our model the (total) demand of region r° accruing to the entrepreneur in r is determined by the relative 'attractiveness' of this entrepreneur to consumers in r°. Consequently, we have:

$$SX_{rt} = \sum_{r^\circ} DX_{r^\circ t} * \left[\frac{A_{rr^\circ t}}{\sum_r A_{rr^\circ t}} \right] \qquad (3)$$

in which:

SX_{rt} = total production of (the entrepreneur located in) region
r at time t.

Thus the above relationship represents a pooling model, in which
the demand attracted to - and in this way the production of - region r
is determined by the relative 'attractiveness' of region r with respect
to all other regions (including r itself) for consumers located in each
type of region.

Next, as regards the ('general') regional *production price* level,
it is assumed that each region (i.e., entrepreneur) starts with the same
uniform production price (p_o), but that afterwards this price will
become lower as the capital-output ratio decreases. In our model, this
ratio will decrease because of technological *'learning economies'*
regarding the related production *process technologies* (i.e., the second
type of further innovative activities stressed in the introduction to
this chapter).[14] In our model, this implies that the productivity of new
capital 'vintages' will be higher than old 'vintages'.[15]

One might also include the influence of labour productivity
(costs) upon the 'general' production price level. However, to avoid
unnecessary complexity and in order to *focus* especially on the *role of*
(the two types of) *technological change* in generating spatial dynamics
(see section 4.1), labour will not explicitly be dealt with in our
analysis. Consequently, we have the following 'general' *production price*
equation for (the entrepreneur located in) region r:

$$P_{rt} = p_o - b * \left[\frac{SX_{rt-1}}{K_{rt-1}} \right] \qquad (4)$$

in which:

p_o = the ('general') production price level at which all
regions start to produce.

K_{rt-1} = total capital stock in region r in period t-1.

b   = 'sensitivity' of the production price level with
respect to the (inverse of the) capital-output ratio.

In this respect, it seems reasonable to assume that
this parameter will be larger the higher the share of
capital costs in total production costs.

Equation (4) includes a one-period time delay from the side of producers
to react to changes in capital productivity.[16]

In the literature, it is usually asserted (cf., Freeman et al
1982, Rothwell and Zegveld 1985) that *production techniques* (of the
variants) of innovation A become *more efficient* in the course of time
(cf., the second type of further innovative activities distinguished in
section 4.1). In this respect, capital 'vintages' often become more
efficient because of (technological) 'learning economies' resulting from
the interaction between producers and capital good suppliers.
Accordingly, we describe the evolution of the capital-output ratio dz_t
of capital 'vintages' as follows:

$$dz_t = g + \left[\frac{d^*}{\sum\limits_{r} \sum\limits_{t=t-1} SX_{rt}} \right] \qquad (5)$$

in which:

dz_t = capital-output ratio (efficiency) of 'capital stock
          vintages' at time t.

g = 'technical minimum' capital-output ratio which will
          be reached in case 'learning economies' regarding
          production process technologies have reached a
          mature stage.

d^* = 'part' of the capital-output ratio that can be
          improved by means of collective 'learning by
          doing'; for this reason the denominator of this
          expression consists of *total* production of *all*
          regions and time periods considered (see also
          below). The larger this sum the less possibilities
          are available for further improving - i.e., decreasing -
          the capital-output ratio.

Consequently, in our model it will be assumed that improved capital 'vintages' are available to all producers. This may be a reasonable assumption as Dosi (1984, pp.147) remarks: 'in the case of capital-embodied (and size-neutral) innovations, there is no a priori impediment to a diffusion of best-practice techniques throughout an industrial sector'.

Thus in our model all regions may benefit (to the same extent) from the improvements in capital efficiency. The question whether and when the (entrepreneurs in the) regions *actually* adopt improved 'vintages', however, will depend on their investment behaviour (see below).

As a consequence of equation (5), we have the following (regional) investment equation:

$$I_{rt} = dz_t * (SX_{rt} - Cap_{rt}), \quad \text{if} \quad SX_{rt} - Cap_{rt} > 0$$
$$= 0, \quad \text{otherwise} \tag{6}$$

in which:

I_{rt} = total investment of (the producer in) region r at time t; as can be seen from the above specification we will *avoid negative investments*.

Cap_{rt} = production capacity of the producer in region r at time t. This capacity is determined by the *investments* from *previous* time periods, the *specific capital-output ratios* of these 'vintages' and the *depreciation rate* (see eq. (7) below).

In our model, further technological improvements concerning the *production process technologies* - and embodied in *(fixed) capital* goods - reflect the *(spatial) inertia* resulting from further innovative activities along the underlying (technological) 'trajectory'.[17] This statement will be further illustrated in the simulation experiments which will follow in the next sections.

Now we have the following relationship for the (regional) *production capacity* (related to former investments):

$$Cap_{rt} = \sum_{t^{\circ}=1..t-1} \left[I_{rt^{\circ}} * (s)^{t-t^{\circ}} \right] / dz_{t}^{\circ} \qquad (7)$$

in which:

s = the general 'survival rate' of capital 'vintages' (i.e., the complement of depreciation)

The *capital goods stock* in region r at time t is now equal to:

$$K_{rt} = \sum_{t^{\circ}=1..t-1} I_{rt^{\circ}} * (s)^{t-t^{\circ}} + I_{rt} \qquad (8)$$

With respect to the number (and quality) of the (further) product-oriented innovations generated in region r, the following relationship is assumed:

$$Prinn_{rt} = \triangle SX_{rt-1} * \left[\frac{C}{\sum\limits_{t=t-1} \sum\limits_{r} SX_{rt}} \right] \text{, if } \triangle SX_{rt-1} > 0$$

$$= 0, \quad \text{otherwise}^{18} \qquad (9)$$

in which:

$\triangle SX_{rt-1}$ = increase in regional production at time t-1

C = a general 'potentiality set' reflecting the overall possibilities for further *improving* the new type of *product* A - i.e., its product characteristics - along the underlying technological 'trajectory' (by means of improved product variants).

Consequently, $Prinn_{rt}$ should be interpreted as a measure of the (rate of) further *product oriented improvements* (i.e., product innovations) a region is able to generate in a certain period t. Ideally, this measure should be interpreted as a weighted sum of the *number* and *quality* of these further improvements (compared to former variants of the new type of product A). For the sake of convenience however, we will subsequently often use the term *number* of product innovations for this combined effect.

As can be derived from the denominator in the above mentioned specification, it is assumed that *especially at the beginning* of the emergence of the new type of product A there is some pool of 'unexploited' possibilities for further improving the product (characteristics). These possibilities will be put in use in the course of the production process of (variants of) such an innovation (cf., Metcalfe 1981, Freeman et al 1982 and Ayres 1987).

In a way the above specification has some similarities with respect to equation (5). As can be derived from the denominator of the present expression, the *possibilities* of introducing *further product improvements* (variants) will *decline* the *more experience* producers[19] acquire with respect to the production of (variants of) innovation A.

Consequently, this specification is in accordance with our prior analysis in the foregoing chapter (see section 3.4). Also with respect to the new type of product itself, *'learning economies'* - resulting in improved product variants - will be generated (although at a decreasing rate). These improvements reflect the *first type* of (further) innovative activities - along the underlying technological 'trajectory'- distinguished in section 4.1.

In our model, demand and innovative activities are *interrelated*. From equation (9) above, it can be derived that the generation of product (oriented) innovations in a region in period t is *positively* related to the increase in demand (i.e., production cf., eq. (3)), accruing to the region in the previous period.[20] *Also*, producers may respond to a *rise* in *demand* by *investing* in more *efficient process technologies* (i.e., capital 'vintages'; see eq. (6)) and *lowering prices* (see eq. (4)). On the other hand, *demand* (see eqs. (2) and (3)) is again *attracted* to those regions which can offer further *product innovations* and *low prices* due to efficient process technologies.

Concerning the 'adoptiveness' parameter a_{rt} from (1), the following relationship is assumed to exist:

$$a_{rt} = (1-\delta_r)*a_o* \left[1 + \frac{P_o - b/g}{\overline{P}_{rt-1}} \right] + \delta_r*\sigma, \quad \text{if } D_r*t > K_c$$

$$= 0, \quad \text{otherwise} \tag{10}$$

in which:

δ_r = 1 in case a region passes the critical threshold K_c (see below) for the *first* time; otherwise δ_r = 0

σ = 'initial' value of a_{rt} in case a region passes the critical threshold for the first time.

$\overline{P}_{rt-1}$ = average price (including transportation costs) paid by consumers (adopters) in region r at time t-1.[21]

$P_o - b/g$ = 'technical minimum' price (if production is taking place by using the most efficient capital 'vintages', i.e., if all 'learning economies' concerning process technologies have reached a mature stage).

K_c = 'critical' threshold level which determines the order in which demand concerning the new type of product A in our distinguished regions 'takes-off'. As stated in the foregoing section, in this respect we will assume a spatial *hierarchical* ordering.

D_r = density of economic activities in region r (r = c, i, p).

a_o = a 'scale' parameter.

In the above specification it is (implicitly) assumed that (potential) adopters have some idea concerning the discrepancy between the actual price ($\overline{P}_{rt-1}$) paid and the technical minimum price (i.e., $P_o - b/g$). Consequently, in case this discrepancy decreases, a_{rt} will increase because more (potential) consumers are willing to join the market, as they will realize that the possibilities of further *price decreases* become *smaller*. In this respect, an increase in the (average) price paid due to transportation costs (i.e., in case consumers are 'shopping' outside their region) has a *downward* impact on a_{rt} and hence on (the growth of) total demand in the region.

In conclusion, the following inferences concerning our model (specification) can be made. Regarding the new type of product A, at the beginning of production possibilities for (further) *product oriented innovations* are (relatively) abundant. Consequently, entrepreneurs will be able to generate many new and improved product variants. In the course of time however, as more and more of these 'learning economies' have been 'reaped', it will become increasingly difficult to generate such types of innovations. In a similar vein, regarding the *production process technologies*, a mechanism of 'learning by doing' will improve the efficiency of these technologies - i.e., increasing the efficiency of capital 'vintages' - although (also) at a decreasing rate.

As regards *demand*, more demand is attracted to a region in case the ('general') *production price is lower* and the number (and quality) of the *product oriented innovations higher*. As to the production price level, this level will decline as *more efficient process technologies*- i.e., capital 'vintages' - are put into operation. Such investments will again depend on the demand attracted to the region. This statement *also* applies to the generation of further product innovations within a region.

Consequently, at a certain moment in time, a region's competitive (production) performance is determined by the three interdependent forces distinguished in the introduction to this chapter:

* the *product* (oriented) *innovations* generated in the region

* structure and age of capital goods ('vintages') determining the *efficiency* of production *process technologies* and, consequently, the ('general') production price in the region

* *location* with respect to (growing) *user markets*, i.e., demand

The foregoing equations make up the ingredients of our aggregate innovation model for illustrating spatio-temporal dynamics related to the three interdependent forces mentioned in section 4.1. In the following sections this model will be further illustrated by means of several computer simulations.

4.4 RESULTS OF COMPUTER SIMULATIONS

In the present section the dynamic innovation model presented in the foregoing section will be used to generate - on the basis of simulation experiments - an 'ideal-type' of spatial dynamics inherent in the emergence and 'take-off' of a new technological 'trajectory'- related to the new type of product A.

Our strategy in the following analysis will be to choose a 'reference' simulation model and to discuss some of the most important results of the simulation carried out. In the next section, some 'sensitivity' experiments will be undertaken in which some of the parameter values of this 'reference' model will be modified.

For our reference model we have used the following 'reference' parameters:

D_c = 100	D_i = 40	D_p = 30	K_c = 300
σ = 0.013	a_o = 0.01	b = 50	
β = 2	s = 1	d^* = 200	
C = 30	P_o = 40	g = 5	
c_r = 1	N = 100		
dci, dic, dip, dpi = 5		dcp, dpc = 10	

In our 'reference' model we will ignore - for the time being - the possibility of depreciation. In our further experiments however (to be discussed in the next section), this possibility will also be considered. In our analysis (see assumption 2 in subsection 4.2.1) the intermediate zone is assumed to be geographically located between the centre and periphery. This is also reflected in the values for the (general) transportation costs between the regions. Now we will present some numerical results.

Figure 4.3 below depicts the *aggregate regional production* at several time periods.

Figure 4.3   Regional Patterns of Production

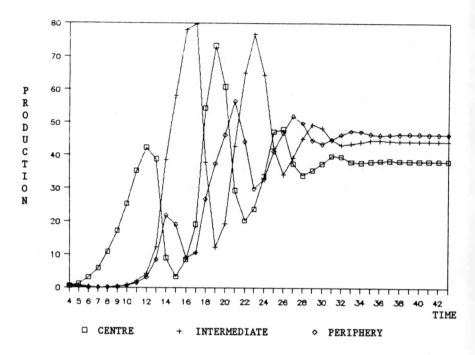

In this respect three time intervals or phases (cf., the phases distinguished in our 'dynamic incubation' framework; see section 3.5) can roughly be distinguished:

A   Incubation phase

This phase refers to the *initial periods* - i.e., roughly consisting of the periods 4-13 - in which the new type of product is first (and especially) *adopted in the centre.*

As regards *production* the centre clearly takes the lead in this phase. This does not only hold for production, but also for the related further *product* oriented *innovations* and *investments* in more efficient *process* technologies as can be seen from the Figures 4.6 and 4.7 below, respectively. In this phase, the *favourable location as regards growing consumer markets* (see Figure 4.4), the (resulting) *product innovations generated* and the *low production prices*[22] (see Figure 4.5) foster the 'attractiveness' of this region (also to 'late' consumer adopters in the intermediate and peripheral zones).

B Competition phase

B1 **Early competition:** This phase rougly covers the periods 14-20
in which the producers in the intermediate and peripheral zones
become increasingly engaged in the production activities related
to the new 'trajectory' and start competing with the (entrepreneur
located in the) central region.

B2 **Late competition:** This phase covers roughly the periods 21-30
in which competition between the regions, on the basis of
location, efficiency of production process technologies and
product oriented innovations is becoming established (leading to
a more balanced situation).

As stated above, it can be inferred from Figure 4.3 that the
centre clearly takes the lead in the production of (variants of)
innovation A. In the competition phase however, the other regions also
become increasingly engaged in the production related to the new type of
product A. This is, inter alia, caused by the fact that in this phase
market opportunities - i.e., aggregate *demand* - expand rather rapidly
in the intermediate and peripheral zones compared to the centre where
market saturation is first approached (see Figure 4.4).

Also, the rather large *production price discrepancies* (see Figure
4.5) - caused by the heterogeneous 'vintage' structures in the regions-
and the *varying degrees* of (further) *product innovations* generated in
the regions (see Figure 4.6) play their roles in the spatial shifts of
production.

Consequently, these two forces illustrate the role of
technological change - i.e., 'creative diffusion' - in generating
spatial dynamics. Although the evolution in the centre - both concerning
demand and production - clearly preceded developments elsewhere
technological change is not a static phenomenon. Further improvements-
both regarding the new type of product itself and the related production
processes (although both at a decreasing rate) - will be generated as
the innovation 'diffuses' through the economic system.[23] These
improvements form an important *cause* of the *spatial shifts of production*

in later phases of this 'creative diffusion' process (i.e., in later phases of the life cycles of industries and technologies).

Concerning *process technologies*, for example, these further innovative activities imply that the intermediate and (especially) the peripheral zones are able to adopt (on average) the more competitive production techniques - and outstrip in this way the centre as the prime producer - in later stages of the competition and stagnation phase. Near the end of the competition phase it can be observed from Figure 4.3 that the fluctuations in regional production levels diminish.

C Stagnation phase

In this phase the consumer market is 'divided' among the three competing regions (entrepreneurs) leading to an 'equilibrium' situation.

In the *stagnation phase* (i.e., roughly after period 30) the centre - although being the place where 'it all started' - becomes the smallest producer. This is due to the fact that in this phase it has 'to pay the price' (see Figure 4.5) - in terms of possessing rather old-fashioned process technologies - of being an early market leader. In this phase the three forces causing the (production) *fluctuations* in the foregoing phases become more *stable* because of the following reasons:

a. Possibilities for further improving the new type of product A (i.e., generating product innovations) diminish, as more and more of the initial possibilities open for such improvements have been put to use.

b. Also, possibilities for further improving the related production process technologies - i.e., the efficiency of capital 'vintages' in our model - become smaller because of diminishing returns of 'learning economies' embodied in (the production of) capital goods.

c. In later phases the *growth of demand* for (variants of) the new type of product *stagnates* in *all regions* concerned. In these later phases demand will largely consist of 'replacement' and not of 'expansion' acquisitions anymore.

When these (three) *interdependent forces* tend towards a more stable pattern, production fluctuations among the regions decrease, and finally the market moves toward a 'stationary state'.

Until now we have especially concentrated on the regional *production patterns* generated by our (simulation) model. In the following pages the results for the other variables will briefly be considered.

From Figure 4.4 below it can be derived that the development of regional demand - concerning the new type of product A - corresponds to the (empirically) well-known *S-shaped adoption curve* discussed before. Consequently, these results of the simulation experiments appear to be fairly plausible.

Figure 4.4 Regional Patterns of Demand

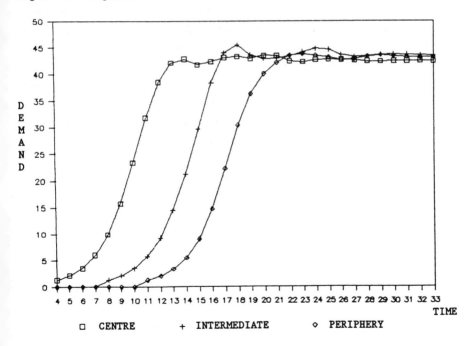

We shall subsequently examine *production price developments* more specifically. It can be inferred from Figure 4.5 below that in the 'incubation' phase the (entrepreneur in the) centre also has a *production price advantage* as compared to the other regions. As stated above, this advantage is caused by the fact that in this phase the entrepreneur located in the centre is investing very heavily in improved capital 'vintages' (i.e., process innovations). Consequently, the *average* efficiency of its production process technologies will *exceed* the efficiency in other regions.

During the competition phase, production prices in the intermediate and peripheral zone fall rapidly below those in the centre, however. This is caused by the fact that the (overall) investment efforts in these regions are *delayed* as compared to the centre. In these later phases, the *capital productivity* - due to 'learning economies' as regards production process technologies (embodied in more efficient capital 'vintages') - has increased compared to the productivity in the 'incubation' phase.[24]

Due to its relatively *low production prices* in the early competition phase (cf., Figure 4.5) as well as its rapidly growing home market (cf., Figure 4.4) the entrepreneur located in the *intermediate zone* attracts a rather large consumer market in this phase. This attraction is further enhanced by the resulting *product innovations* generated in this region (cf., Figure 4.6). As to the entrepreneur located in the *peripheral zone*, production becomes significant during the late competition and stagnation phase. As can be seen from Figure 4.5 below, the periphery is able to offer relatively *low production prices* - because of its efficient *process technologies* - in these later phases. This largely explains its '*attractiveness*' (cf., Figure 4.3) in the later stages.

Figure 4.5   Regional Patterns of Production Prices

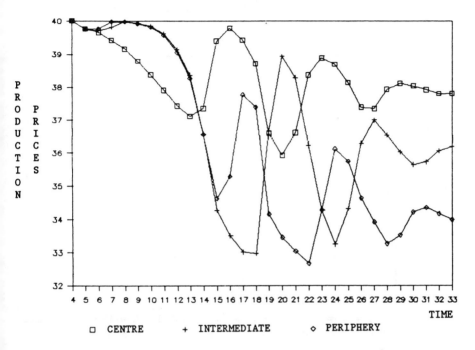

Should we look more specifically at the (further) generation of *product innovations*, it can be derived from Figure 4.6 below that the greater part of these innovations has - especially in the *'incubation'* phase - been *generated* in the *centre*. As stated above, this explains part of the 'attractiveness' of the centre in this phase.

However, it can also be derived from this figure that *not all* these innovations have *exclusively* been generated here. Especially in *later phases* the regional discrepancies level off and also the entrepreneurs located in the intermediate and peripheral zones are able to generate[25] (some of) these innovations.

These observations are in accordance with our theoretical framework discussed in the foregoing chapter. In that context we hypothesized that especially central (metropolitan) regions - in the *initial phases* of the 'take-off' of new technological 'trajectories'- will be important *'incubation' places* of such innovations. In later phases however, such innovations - although at a lesser rate - will also

136

be generated in other non-metropolitan regions.

Figure 4.6   Regional Patterns of Further Product Innovations

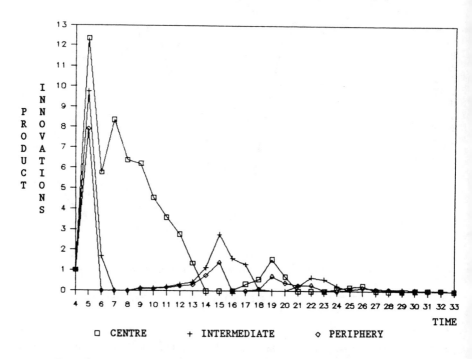

□ CENTRE   + INTERMEDIATE   ◇ PERIPHERY

By taking investments as a proxy for process innovations - i.e., incorporating the most recent (efficient) capital 'vintages' in the production process - we can derive from Figure 4.7 below that *at the beginning* of the 'take-off' of the new 'trajectory' the centre will also adopt the greater part of the new capital goods, i.e., adopt process innovations. Later on, however, it will be *especially* the *intermediate* and *peripheral zones* which will incorporate the further improved process technologies, since their *investment 'boom' is delayed* as compared to the centre. In fact, as can be seen from this figure, the centre acquires the largest share of its capital goods in the initial 'incubation' phase, while the other zones have their investment 'peaks' in the competition phase.

Figure 4.7   Regional Investment Patterns

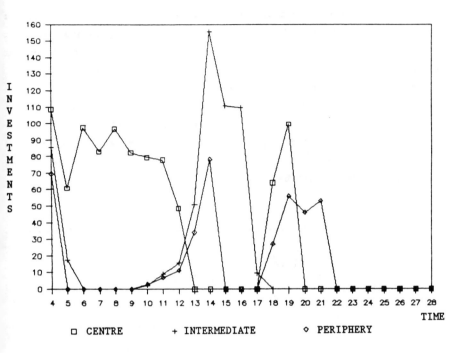

Consequently, as regards the (further) innovative activities especially the centre - i.e., in the 'incubation' phase - can be considered as the innovative heartland. In later phases however, the intermediate and peripheral zones - especially regarding *process innovations* - tend to become increasingly important which *enhances their competitive power*. In this respect, the integration of *ongoing* (endogenous) *technological change* - both regarding the production techniques and the new product (characteristics) itself - with the (spatio-temporal dimension of) user markets will lead to. *spatial fluctuations of production*.[26]

In this section we have illustrated the model developed in section 4.3 by means of computer simulations. As regards the parameter values of this model, we have chosen a 'reference' set. In the next section, a sensitivity test on several parameters of this set will be carried out (one at a time). This may (further) enlarge the understanding of the (conceptual) model developed in section 4.3.

4.5 A SENSITIVITY ANALYSIS

In the present section we will present a number of results of simulation experiments in which we changed one parameter of our 'reference' at a time. In Table 4.1 we present the most important impacts of these parameter changes - related to the 'incubation', competition and stagnation phases - which can be derived in a qualitative sense. In this table we have concentrated on the impacts of such parameter changes on regional *production* and 'final' (i.e., after 40 periods) *production price differences* between the regions distinguished. These impacts - i.e., compared to the outcomes of our 'reference' model - will be further discussed below.

Table 4.1 Impact of Changing Parameter Values of the 'Reference' Model

Phase	Region	Parameter Change						
		$g=10$	$s=0.97$	$d^*=350$	$K_c=225$	$dci=7.5$ $dcp=15$	$C=50$	$\beta=3$
incuba- tion	c	+	0	-0	+	0	+	+
	i	0	0	0	0	-0	-0	-0
	p	0	0	0	0	-0	-0	-0
early comp.	c	+	0	0	0	0	increase	
	i	+	0	-0	+	0		
	p	0	0	0	+	-0	in	
								idem
late comp.	c	+	0	0	+	0		
	i	+0	+0	+0	-0	+0		
	p	0	+0	-0	0	-0	fluctu-	
stagna- tion	c	++	++	0	0	-0		
	i	+	+	0	-	-0		
	p	0	-	0	+	-0	ations	
producti- on prices at period 40	c	38.5	34.3	38.5	38.0	37.9	38.0	37.7
	i	37.8	32.0	36.9	36.7	36.2	36.7	36.8
	p	36.5	31.3	34.8	33.5	33.8	34.7	36.4

Legend:
+ = positive impact on production compared to 'reference' model
0 = idem neglectable impact
- = idem negative impact

An *increase of g* - i.e., the 'technical minimum' capital-output ratio - implies that the (technically) minimum production price (p_o-b/g) *rises* compared to our 'reference' model (as can be seen from eq. (4)). As a consequence, the (technically) *possible price decreases* become *smaller*.[27] In this respect, it will become less attractive for (potential) consumers to postpone their adoption. Consequently, an increase of g (to a level of 10) in our 'reference' model will result in a faster (initial) growth[28] - and final level(s) - of (regional) demand.[29]

Also, and accordingly, *production price discrepancies* between the regions - resulting from regional differences in the efficiency of process technologies (i.e., 'vintage' structures) - will *level off*. In this respect, in a *relative sense* transportation costs (and product oriented innovations) become more important in determining the 'attractiveness' of the respective regions to consumers.

As can be derived from Table 4.1 above, this parameter change appears to be especially *'beneficial'* to the *centre*. In this case, the old-fashioned process technologies of the centre are less important in generating relatively high *production prices* in the competition and stagnation phases. Also the intermediate zone 'benefits' from this parameter change as - in a relative sense - location (in the midst of consumer markets) becomes more important in determining regional 'attractiveness'.

The opposite conclusions hold for the peripheral zone however, as its competitive performance - resulting from a favourable 'vintage' structure in later phases - decreases. Consequently, this region is unable to attract any significant proportion of the increases in the regional levels of demand.

Hitherto, we have abandoned the possibility of *depreciation*. Should we introduce a 'survival rate' of capital vintages of 0.97, the overall *production changes* during the first phases are only marginal. In the course of time, the regional (production) impacts become more interesting, however. In these later phases the regional levels of *demand* will *increase* - compared to the situation of non-depreciation-as then the technically minimum production price will be approached.[30] Also, in the course of time, regional *production price discrepancies*-resulting from heterogeneous 'vintage' structures - will *level off* because of depreciation of old inefficient 'vintages'.

This implies, for example, that in the stagnation phase the *central region* has *improved* its *relative* production (price) performance as compared to the 'reference' model. The opposite conclusion holds for (the entrepreneur located in) the peripheral region, however. In these later phases also the entrepreneur located in the intermediate zone is increasingly capable of reaping the fruits from his favourable geographical location.

In the *limit*, regional production price discrepancies will even *disappear* as then all regions will apply the same - technically most efficient - 'vintages'. As Table 4.1 suggest, however, this may indeed be a long-term process.[31] Such a 'stable' situation will be approached *sooner in time* the *lower the 'survival rate'* of capital 'vintages'. However, such a 'stable' limit situation may very often never be reached as - in the course of time - the markets of the prevailing 'new' type of product will (again) be *invaded by new technological 'trajectories'*. We will return to this issue in the conclusion of this chapter.

Changing the production process *'learning parameter' d^** (to a level of 350) appears to have only marginal impacts (i.e., compared to the results of our 'reference' model). In the *initial phases* this change in parameter value will have a downward impact on regional demand levels as the (average) prices paid will *diverge more* from the 'technical minimum' production price. In this respect, it becomes more attractive for (potential) consumers to *delay* their *adoption patterns*. Consequently, in the 'incubation' phase this will (somewhat) decrease regional production levels in the centre.

Later on, when these 'learning economies' have become more mature, this impact will diminish. Concomitantly, there is a tendency that regional *fluctuations* in production levels - caused by the overall delayed pattern of adoption by potential consumers - will be *more pronounced* in the (late) competition phase than in our 'reference' model.

In the stagnation phase *overall production prices* will be somewhat *higher* as - especially in the initial phases of the regional investments - the efficiency of capital is rather *poor* compared to our 'reference' model. The regional *discrepancies* in these prices (and production) will differ only marginally from those in the 'reference' model in this phase, however.

In case the regional *patterns of adoption* by (potential) consumers become more *homogeneous*[32] - i.e., in case the time-lag between the 'take-off' of demand in the regions considered becomes smaller (i.e., lowering K_c to a level of 225) - regional *fluctuations* in production become more *pronounced* in the 'incubation' and early competition phase. Consequently, in these first phases, regions exhibit larger production 'peaks' - i.e., attract more demand - as compared to our original model. On the other hand, these fluctuations *level off* during later phases as the regional adoption curves become more homogeneous.[33]

In the stagnation phase regional *production price discrepancies*-especially between the intermediate and peripheral zone - have increased.[34] For this reason, the latter zone attracts a larger share of the market - at the expense of the intermediate zone - in the stagnation phase as compared to our 'reference' model.

As discussed in the foregoing section, increasing *transportation costs* (to a level of 7.5 and 15, respectively)[35] will have a *negative impact* on aggregate regional *demand* as the speed and rate of adoption will level off (resulting from lower values for the 'adoptiveness' parameters a_{rt}: see equation (10)). The influence on - i.e., changes in - regional patterns of *competiveness* appears to be only marginal in this particular case, however. The negative impact of decreased adoption in the different phases on regional production levels applies quite uniformly to all regions concerned. Also the regional *differences* in the *production prices* in the stagnation phases are (more or less) *similar* to those in the 'reference' model.

Both the impact of increasing C (i.e., possibilities for generating further product innovations) and β (e.g., increasing rationality or unanimity among users) *reinforces* regional *fluctuations* in production levels as might be expected a priori. When C increases, for example, the possibilities for further improving the new type of product itself - following and resulting in demand increases - will be higher. Consequently, the 'hills' and 'valleys' of regional production patterns will become more *pronounced*. The same conclusion applies to an increase in β, as also in this case a greater proportion of (regional) *demand* will shift towards the *same* (most 'attractive') region at a certain moment of time.

In the stagnation phase an increase in C (to a level of 50) *reduces* regional *production price discrepancies*, especially between the centre and the periphery. This reduction is even larger in case β is changed to a level of 3. As in both these cases *all* regions experience higher production 'peaks' in the first (two) phases, they will *invest more* in (by then) rather *inefficient* process technologies. Consequently, part of the 'success' of the intermediate and (especially) the peripheral zone in the later phases[36] - resulting from being relatively 'late' investors - is being 'destroyed' in this way.

In this section we have discussed, in a *qualitative* sense, the *partial* impacts of changing one parameter at a time in our 'reference' model. The overall patterns observed in the foregoing section appear to be quite robust, although specific patterns will of course differ somewhat. In the next section a number of conclusions, based on the foregoing analyses, will be presented.

4.6 CONCLUSION

In the foregoing sections we have described an 'ideal-typical' pattern of spatial dynamics resulting from the emergence - and further innovative activities along - a new technological 'trajectory' related to a new type of product A. We shall now assume that *several* new 'trajectories' will be generated according to a time-pattern as depicted in Figure 4.1. In case (a large share of) these new 'trajectories' display(s) a *similar pattern of spatial dynamics* as our ideal-typical 'trajectory' considered in the previous sections, the resulting regional impacts can be traced by means of 'aggregating' our foregoing analysis. This would, for example, imply that the *'clustering'* of the emergence and 'take-off' of new 'trajectories' also has its spatial counterpart leading to new *spatial* life cycles of *industries and technologies*.

If we would further *extend* the (qualitative) analysis of this chapter - as regards a *specific* new technological 'trajectory' - and also consider the possibility and spatial consequences of new 'trajectories' *invading* the markets of former 'trajectories' (cf.,

Batten 1982), the following final remarks are appropriate here. On the basis of our framework (model) this 'invasion' process can again be expected to *start from the centre* onwards. In this respect, *changing consumer/user behaviour* (adoption), for example, can - in the similar way we assumed with respect to our new type of product A - *first* expected to be noticed in the more central (metropolitan) areas.

Consequently, it follows from the same argument that new technologies are generally expected to be first adopted in the central regions, that old technologies - in as far as they will be 'substituted' by the new technologies - will *first be 'demolished' here.*[37]

In this way, the central regions would *precede* other regions in a (dis)continuous *structural transformation process* in which 'new technologies' compete with 'old technologies' (see also the first hypothesis deduced from our framework in subsection 3.6.1). This 'take-over' process may result in a 'buy-out' process in which firms linked to such new 'trajectories' - which especially in the beginning will generate high values added - can offer the highest bid-rents for the most favourable (central) locations. This would further depress the competitive performance of firms linked to the former 'trajectories' in this region.

In this chapter, the 'dynamic incubation' framework developed in Chapter 3 has been illustrated by means of a behavioural model for spatial innovation patterns. In this respect, the role of technological change - i.e, the emergence and 'take off' of new technological 'trajectories' and the 'creative diffusion' process along these 'trajectories' - in generating spatial dynamics has been analyzed.

Our model (framework) takes the central (metropolitan) region as the starting point where such new 'trajectories' - and the resulting new life cycles of industries and technologies - will *first* be adopted and the economic impacts will be observed. In our (illustrative) model discussed in this chapter, we have taken the spatial hierarchical development pattern of demand as the main 'driving force' of this initial leading position. As stated before, other forces - e.g., the existence of *production agglomeration economies* in central regions in the initial phases of production (cf., Hoover and Vernon 1959 and Carlino 1978) - might equally well apply here.

In our illustrative model, the role of the *'creative diffusion'* process - i.e., the generation of further product and process innovations along new 'trajectories' - in generating spatial fluctuations in production has also been analyzed. In this context, it has been demonstrated that this process may lead to spatial deconcentration in later phases of life cycles of industries and technologies. Consequently, in our (model) framework technological change itself is an important (endogenous) driving force of spatial dynamics.

As stated in the previous chapter (see subsection 3.6.1) the lack of relevant data - e.g., concerning the space-time trajectory of new industries,[38] the spatio-temporal dimension of innovative activities in such industries - precludes an in-depth and 'full-scale' empirical test of this framework. However, several aspects of this framework are embodied in - or confirmed by - various spatial innovation research efforts discussed in Chapter 2 of this study. Besides and above this, in subsection 3.6.1 of Chapter 3, several testable hypotheses have been deduced from this framework. These hypotheses will be analyzed in the second empirical part of this study. Consequently, these investigations will also provide - modest and partial - tests concerning the empirical validity of our 'dynamic incubation' framework developed in the previous chapter and further illustrated in the present chapter.

This chapter concludes the mainly theoretical and descriptive parts of this study. In the second - i.e., empirical - part of this study the first basic research issue on the impact of both the 'structural' and 'production milieu' component on spatial variations in the innovative performance of firms will be considered. As stated above, in this second part an attempt will also be made to examine the empirical validity of the theoretical framework - developed in the context of the second basic research issue identified in Chapter 2 - by means of testing the hypotheses derived from this framework in the previous chapter. In the next chapter, we will start with the first hypothesis on the leading role of metropolitan or central areas concerning the growth and decline of life cycles of industries and technologies.

Notes to Chapter 4:

1. For the sake of convenience, in the following pages we will mainly refer to a new type of *product*. However, the same kind of mechanisms may also apply to new types of *services*.

2. In this respect, the spatial patterns of production, further innovative activities, demand and the like will be considered.

3. In this context, we will consider a 'trajectory' which is not (permanently) dominated by one specific firm (see also the previous chapter).

4. Consequently, in the following pages we will often refer to the production, innovations of a *region* instead of an *entrepreneur*.

5. 'Consumers' may refer to 'households' or 'firms' (or both) depending on whether the innovation concerned is a final or intermediate good (service).

6. From the foregoing chapters it will be clear that such a spatial adoption pattern of new economic activities is well-established in spatial innovation analysis. As to the causes of such a pattern several factors - e.g., availability of social overhead capital, information flows, high-income levels in urban areas and so on - can be gathered from the literature.

7. As stated before, our model presented below may equally well apply to intermediate goods or even services.

8. See also the previous chapter.

9. In this respect, we will (implicitly) assume that each adopter repeats its adoption decision every time period as the 'duration of life' of the final product will be assumed to last only *one period of time*.

10. See assumption 4 in subsection 4.2.1.

11. Consequently, a kind of 'sequential choice model' underlies the (assumptions inherent in the) first two equations. In this respect, (potential) consumers *first* decide whether or not to adopt the new type of product in period t. In a *second-round* of decision-making, questions such as the choice of product variant and where to buy enter the scene.

12. This is the price, excluding transportation costs, asked by the producer.

13. This is assumed in our analysis; see assumption 3 in subsection 4.2.1.

14. In this respect, it is (implicitly) assumed that the capital 'vintages' are *not* variant-specific. Consequently, with the same piece of capital it is possible to produce - without any additional costs- the various (improved) variants of innovation A.

15. In this respect, capital may - besides machinery and building- also include the mere social overhead capital (i.e., infrastructure) necessary to produce (variants of) innovation A in a region (cf., Hirschman 1958 and Nijkamp 1983, 1986).

16. This may be a reasonable assumption, given the fact that normally prices do not respond immediately to changes in input prices.

17. See also the sixth issue in section 3.4 of the previous chapter.

18. In our simulation experiments $Prinn_{rt}$ has been set equal to 1 in each region in case production starts, i.e., in case demand 'takes- off' in the central region.

19. This experience is reflected in the denominator of this equation.

20. For example, because of the inherent possibilities for increasing R&D efforts.

21. The average price paid in region r at time t equals:

$$\bar{p}_{rt} = \sum_{r^\circ} \left[DX_{rt} * \frac{A_{r^\circ rt}}{\sum_{r^1} A_{r^1 rt}} * (p_{r^\circ t} + d_{rr^\circ}) \right] / DX_{rt}, \quad r^\circ, r^1 = c,i,p$$

22. Resulting from the investments in more efficient process technologies.

23. Or, stated in terms of chapter 1, diffusion cannot be considered as a 'carbon-copy' process of imitation.

24. In which the producer in the central region has been investing rather heavily, cf., Figure 4.7.

25. Although the 'overall' possibilities for generating such innovations will decrease in the later phases (cf., Figure 4.6).

26. Note the fact that in our model the spatial dynamic patterns of production have been generated without having to 'rely' on a- technological factors such as regional discrepancies in wage rates, congestion effects, pollution regulations and unionization. On the contrary, in our model these patterns are largely depended on the role of technological change itself.

27. Note that this statement is valid, ceteris paribus the other parameter values.

28. In this respect, it becomes less attractive for potential consumers to delay their adoption.

29. An increase in g, *ceteris paribus* the other parameter values of our reference model, implies that the 'adoptiveness' parameters a_{rt} will increase (see equation (10)).

30. This will result in higher values of a_{rt} as compared to our reference model (see eq. (10)).

31. Even after 50 periods production price discrepancies are still considerable (i.e., 34.1 in the centre, 31.6 in the intermediate zone and 31.1 in the periphery).

32. For example, in the case less 'complex' innovations are considered.

33. In terms of Figure 4.4 the curves of the intermediate and peripheral zones will shift to the left.

34. In this respect, especially the entrepreneur located in the intermediate zone has to 'pay its price' for his accelerated - compared to the original model - investment efforts in inefficient process technologies in the initial phases.

35. This implies increasing dci, dic, dip and dpi to a level of 7.5 and dcp and dpc to a level of 15.

36. In terms of possessing efficient production process technologies.

37. As stated before, our framework is concerned with the ideal-typical spatial patterns of life cycles of industries and technologies. Clearly, the availability of appropriate incubator facilities such as knowledge centres and science parks in specific peripheral areas, for example, may induce a stronger or a leading position of such areas as regards specific new technologies (cf., the impact of the University of Technology in Twente upon micro-electronics developments).

38. See also Chapter 9 of this study.

PART B
EMPIRICAL
INVESTIGATIONS

5 Structural changes in the Amsterdam area

5.1 INTRODUCTION

The first part of our study (Chapters 1-4) has focused attention on the incubation potential of areas seen from the perspective of technological innovation. Based on a broad literature survey and on various simulation modelling experiments it was regarded plausible that metropolitan or central regions provide especially a seedbed function for new life cycles of industries and technologies, whilst in later phases the role of less centrally located regions may become increasingly important.

The present part of our study addresses the issue of the empirical validity of the above mentioned dynamic incubation framework. By way of illustration we will begin our practical investigations with a brief sketch of (historical) trends in the Amsterdam region. This area plays an important economic role in the so-called Randstad of Holland, the economic motor of the Netherlands (see Hall, 1984). The development of Amsterdam has in this century exhibited various fluctuations which linked to the economic and technological restructuring of the country as a whole. At times there have been serious concerns about the economic performance of Amsterdam, but the city has always arisen like a 'phoenix from the ash'. Therefore, it may be interesting to look at the evolution of Amsterdam from the viewpoint of 'long-term sustainable development' (cf., Ewers and Nijkamp 1989, Davelaar and Nijkamp 1989f).[1]

Consequently, in this chapter some broad (historical) patterns concerning structural (economic) transformations in the Amsterdam region will be outlined. In this context, we will particularly concentrate on the question whether structural changes in this region precede development patterns elsewhere in the Netherlands.

This research issue is also important for the empirical testing of our theoretical framework developed in the first part of this book. On the basis of this framework we expect - see in particular the first hypothesis deduced from this framework in Chapter 3 - that new 'technology systems' (or the impact of these systems) are first realized in large centrally located metropolitan areas like the Amsterdam area. On the other hand, it is also expected that later on the indigenous dynamism will shift to less central (metropolitan) areas.

Consequently, on the basis of the first hypothesis mentioned above, it is expected that the structural decline in later life cycles of industries and technologies will also first be noticed in such regions. In this respect, we have more in particular called attention to the role of innovative activities in generating such a spatial deconcentration trend in later life cycles of industries and technologies.

Consequently, in order to test the empirical validity of the above mentioned theoretical framework, in section 5.2 of this chapter a number of (historical) trends regarding the relative performance of the Amsterdam region with respect to several (types of) manufacturing sectors will be considered. In this section, it will be demonstrated that in recent decades the Amsterdam area had a rather poor performance - also in relative terms - as regards the development of employment in the manufacturing sector (see also Jobse and Needham, 1988). Rather similar conclusions apply to the other central (metropolitan) regions in the Netherlands.

As a follow-up of the analysis in section 5.2, Chapters 6 to 8 of this book will analyze in more detail (based on industrial surveys) whether these employment trends are also mirrored by the innovation patterns - as stressed in our theoretical framework - in these regions. In Chapter 7, for example, the relative performance of the Amsterdam region with respect to the indigenous innovation potential of several types of manufacturing firms will be considered.

In section 5.3 also the relative (historical) performance of the Amsterdam region with respect to (employment in) several types of service sectors will briefly be discussed. For these sectors it will also be analyzed whether the Amsterdam area can be considered as a predecessor of structural developments taking place elsewhere in the Netherlands. However, also with respect to the service sector, a more in-depth analysis of the relative performance of this region - in particular with respect to the producer service sector - is necessary; this will be undertaken more substantially in Chapter 9 of this book. Section 5.4 will contain a number of retrospective and prospective concluding remarks.

5.2 THE MANUFACTURING SECTOR IN THE AMSTERDAM AREA

During the period from 1880 until World War II the city of Amsterdam appeared to be rather successful in attracting a considerable part of the, by then, most innovative manufacturing sectors. At this time Amsterdam provided the seedbed of many new manufacturing activities (cf., Nozeman 1980). Although data limitations restrict very severely the possibilities for an in-depth quantitative analysis, Nozeman estimates that between 1899 and 1909 Amsterdam succeeded in attracting about 25% of the total growth in manufacturing employment in the Netherlands as a whole. At that time Amsterdam's share in total population amounted to approximately 10%. Until 1930 Amsterdam increased its *share* in total manufacturing employment (see Nozeman 1980).

Although afterwards its share in manufacturing employment steadily decreased, it is generally acknowledged (cf., Nozeman 1980, Jansen and de Smidt 1974, Mouwen 1984) that with respect to the most expansive and innovative manufacturing sectors Amsterdam still (i.e., until 1950) remained one of the leading cities in the Netherlands. This is demonstrated, for example, by the growth of employment in the chemical and metal industry in Amsterdam which tripled in the period 1907 to 1947 (see Nozeman 1980).

Thus, in general there is no doubt that before World War II Amsterdam performed an important seedbeed function with respect to the (more innovative parts of the) manufacturing sector. A favourable railway and water infrastructure coupled with a large and growing local consumption market appeared to be a strong locational factor.

Soon after World War II however, Amsterdam rather rapidly lost

154

its manufacturing base. Even in *absolute* terms employment in the
manufacturing sector decreased rather drastically as can be seen from
Figures 5.1 and 5.2 below.

Figure 5.1 Development of Manufacturing Employment in Amsterdam in
the Period 1950-1978

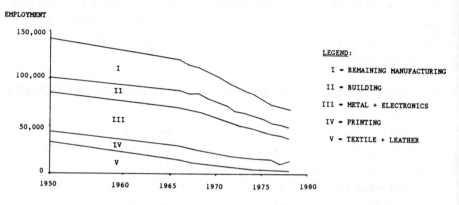

Source: Nozeman 1980, pp.164

Figure 5.2 Employment in some Manufacturing Sectors in Amsterdam
1970-1982

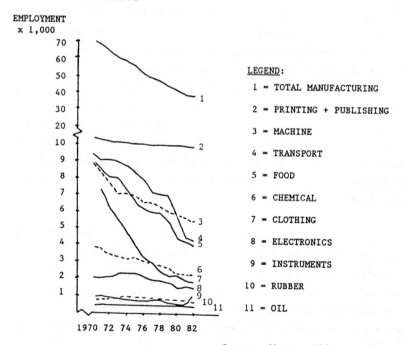

Source: Mouwen 1984, pp.81

Figures 5.1 and 5.2 above clearly demonstrate the absolute decline of employment in several manufacturing sectors in the city of Amsterdam. Compared to the Netherlands as a whole however, the development of Amsterdam clearly *preceded* the structural decline - i.e., in terms of employment - of the manufacturing sector. This is illustrated in Table 5.1 below.

Table 5.1 Manufacturing Employment in the Netherlands*

Year	1953	1958	1963	1968	1971
Employment	862,700	969,500	1068,400	1053,200	1047,200
Index employment	100	112	124	122	121

*excluding mining, including public facilities

Source: Jansen et al 1974, pp.41

These results indicate that between 1953 and 1963, for the Netherlands as a whole, growth of manufacturing employment still exceeded 2% per year. In the sixties industrial employment more or less stabilized, while in the seventies decline started for the Netherlands as a whole (see Davelaar and Nijkamp 1989a and Chapter 9 of this book). Thus as a general conclusion we can state that the decline of the manufacturing sector in Amsterdam clearly preceded the decline in the Netherlands as a whole.

Looking more in depth however, and dividing the total manufacturing sector in a more or less innovative group (based on R&D intensities), Table 5.2 illustrates that especially with respect to the 'old line' sectors - i.e., sectors related to former technological 'trajectories' - the relative position of Amsterdam appeared to be very poor in 1970. With respect to the more innovative parts on the other hand, Amsterdam's position was far less unfavourable (this confirms the above mentioned pattern until 1950).

Table 5.2   Manufacturing Sectors in Amsterdam in 1970

		% of establish-ments located in A'dam 1970	idem with respect to employment
More Innovative Sectors	instruments	15.4	13.7
	chemical	12.6	7.1
	rubber	10.3	4.6
	machine/engineering	10.3	11.7
	petro-chemical	9.5	2.1
	electronics	11.0	3.7
Less Innovative Sectors	food	4.7	5.7
	building materials	7.9	4.3
	wood and furniture	5.9	1.9
	textile	3.4	0.6
	total manufacturing	8.9	7.3

Source: Mouwen 1984, pp.76-77

Thus in 1970 the relative position of Amsterdam appeared to be (far) less worse (or even favourable) with respect to those sectors which can be considered to be more closely related to more recent 'technology systems' than those sectors operating under a more or less standardized technological regime. After 1970 however, also the situation with respect to these more innovative sectors (further) deteriorated. Table 5.3 below summarizes a shift and share analysis by de Jong and Paap (1983) concerning the development of establishments in the chemical and electronical sectors in Amsterdam compared to the Netherlands as a whole.

As the results with respect to the location factors in Table 5.3 indicate, developments in chemistry and electronics after 1968 clearly depict a relatively weakening position of Amsterdam.

Table 5.3 Location Factors of a Shift-Share Analysis with respect to
Establishments in Chemistry and Electronics

| | Location Factor* Amsterdam | |
Year	Chemistry	Electronics
1969	-8.45	-7.9
1970	-2.16	-6.51
1971	-6.49	7.41
1972	-2.88	-4.53
1973	-1.41	-12.1
1974	-6.45	-6.71
1975	-0.77	-6.03
1976	-1.69	-5.35
1977	-5.97	-7.05
1978	-4.88	-1.25
1979	-1.40	-1.80
1980	-3.11	-1.47
1981	-3.80	

* Related to the shift-share analysis

Source: de Jong and Paap 1983, pp.28

More recent trends as regards the spatial development of
employment in the manufacturing sector in the Netherlands are presented
in Table 5.4 below. On the basis of the Statistics of Employed Persons
we calculated the regional shares in total manufacturing employment for
the years 1973 and 1986. Besides the performance of Amsterdam the other
two large metropolitan areas - i.e., Rijnmond and The Hague - have been
considered separately. In this table the performance of these
metropolitan regions has been compared with other (types of) regions.

In this context, we subdivided the Netherlands in a so-called central
(i.e., the Rimcity), intermediate (or 'halfway') and peripheral zone.[2]

158

Table 5.4 Regional Shares in Total Manufacturing Employment

zone	regional share of manufacturing employment in 1973	idem with respect to 1986	change in share 1973-1986
CENTRAL:			
Amsterdam	5.4	3.6	-1.8
The Hague	2.8	2.5	-0.3
Rijnmond	8.7	7.4	-1.3
Remaining central	20.7	20.1	-0.6
INTERMEDIATE	34.7	37.8	+3.1
PERIPHERAL	28.0	28.6	+0.6

Calculations based on:
Statistics of Employed
Persons 1973 and 1986

As can be derived from this table, besides Amsterdam also the other large metropolitan areas Rijnmond and The Hague have a relatively poor performance as far as employment in the manufacturing sector is concerned. In this context, also the share of manufacturing employment in the other (remaining) regions located in the central zone of the Netherlands has decreased.[3] Consequently, these patterns for the Dutch context appear to be largely consistent with the spatial shift of manufacturing employment as observed in several EEC-countries (cf., Keeble et al 1983). In Chapter 7 of this book it will be analyzed more in-depth whether these employment trends are also reflected by innovation patterns (as stressed in our theoretical framework).

On the other hand however, the peripheral and (especially) the intermediate zone have increased their shares in manufacturing employment in the period 1973-1986. As far as the intermediate zone is concerned, the only region in which the share of employment in manufacturing decreased rather considerable is the (COROP) region Arnhem-Nijmegen.[4] This region is also largely 'metropolitan' in nature (which confirms the above mentioned patterns).

At the regional level of provinces, especially the southern parts of the Netherlands have increased their regional shares in manufacturing employment rather considerably (see also Wever 1985). The province of Brabant increased its share from 19.0% in 1973 to 20.8% in 1986, while in the same period the province of Limburg increased its share from 8.6% in 1973 to 9.7% in 1986. In Chapter 7 it will be demonstrated that the favourable employment patterns in these provinces also appears to be reflected by rather favourable innovation patterns (e.g., concerning the 'new line' firms).

After we have analyzed the relative performance of the Amsterdam region regarding the development of employment in the manufacturing sector, we will subsequently analyze the relative performance of this region with respect to several types of service sectors.

5.3 SERVICE SECTORS IN THE AMSTERDAM REGION

Concerning employment in the service sector as a whole, Amsterdam experienced a 'limit to growth' earlier than the Netherlands as a whole. Employment within the total service sector reached its highest level in 1970 in Amsterdam after which growth stagnated or even became negative as can be derived from Figure 5.3 and Table 5.5 below.

Table 5.5 Employment in the (Total) Service Sector in Amsterdam

Year	1975	1977	1979	1981	1982
Employment in service sector in Amsterdam	229,403	220,125	216,574	217,455	213,355

Source: Bureau of Statistics of Amsterdam (Amsterdam in Cijfers, several volumes).

Figure 5.3 Development of Service Sector Employment in Amsterdam
1966-1978

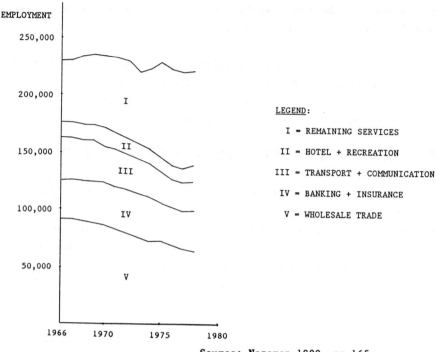

Source: Nozeman 1980, pp.165

For the Netherlands as a whole employment in the service sector still expanded in the seventies, while only in the eighties stagnation - especially because of the decline in the building services - became more apparent (see Buursink, 1985).

In a way analogous to the manufacturing sector, we will now discuss more in depth the developments *within* the service sectors. It is nowadays increasingly acknowledged that *information* is becoming one of the most important production factors (cf., Naisbitt 1984, Ewers 1986, Giaoutzi et al 1988). In this context, the proportion of knowledge inputs in the production processes is rapidly gaining momentum (cf., De Haan and Tordoir 1986). In this respect, sectors related to the more recent information 'technology system' are rapidly expanding (cf., Naisbitt 1984, Freeman 1987a, 1987b; see also Chapter 9 of this book).

Some of the most important sectors in this respect appear to be finance, insurance and (especially) producers service activities such as computer services, technical advice and design, management consultancy (cf., Noyelle et al 1984, Daniels 1985, Marshall 1985). These sectors will be denoted as the FIPS-sector in the remaining part of this chapter.

The relative performance of the Amsterdam region with respect to these sectors will only briefly be considered. In Chapter 9 we will provide a more in-depth analysis of the spatial development pattern of some (new and innovative) producer service sectors related to the more recent information 'technology system'. In that chapter the relative performance of the Amsterdam region with respect to these sectors will also be analyzed in more detail.

Because of data limitations a time series analysis of the spatial development pattern of (employment in) this so-called FIPS sector is impossible, therefore we have to confine our analysis to some basic years.[5]

As Table 5.6 below illustrates, by 1963 Amsterdam possessed a relatively strong position as regards employment in the FIPS-sector. By then about 25% of the FIPS-employment was located in Amsterdam.

Although some care should be taken in comparing the results of 1963 and 1973 (because definitions have changed), it can be derived from Table 5.6 below that also in the period 1963-1973 the FIPS-sector expanded in Amsterdam. In this respect, Amsterdam still 'attracted' about 16% of the total growth of employment in the FIPS sector. By 1973 none of the other large cities appeared to be so heavily biased towards the FIPS-sector as Amsterdam.

Table 5.6   Employment in the FIPS sector

Year	Netherlands	Amsterdam	Rotterdam	The Hague	Utrecht
1963[1]:					
employment in FIPS	192,240	46,070	23,842	29,714	8,259
% of employment (within region) in FIPS	5.6%	13.2%	7.8%	15.8%	9.5%
location quotient[6] employment FIPS		2.35	1.4	2.8	1.7
1973[2]:					
employment in FIPS	284,100	60,820	31,790	29,960	14,560
% of employment (within region) in FIPS	7.4%	18.1%	11.1%	15.2%	12.8%
location quotient employment FIPS		2.5	1.5	2.0	1.7

Calculated from : 1) <u>Regionaal Statistisch Zakboek</u> 1972, pp.73
: 2) <u>Regionaal Statistisch Zakboek</u> 1974, pp.71

As Figure 5.4 and Table 5.7 below reveal, the FIPS sector in Amsterdam has grown rather rapidly in the sixties and early seventies, while its growth levelled off in the late seventies and early eighties. In the period 1975-1980 Amsterdam attracted only few per cent of total growth of employment in the FIPS sector, whilst in this period employment in this sector expanded rather drastically in the Netherlands as a whole.

Figure 5.4   FIPS-sector in Amsterdam 1964-1973

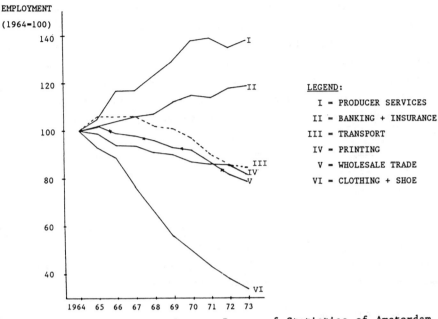

Source: Bureau of Statistics of Amsterdam
(<u>Amsterdam in Cijfers</u> 1975, pp.124)

Table 5.7   Employment in FIPS in Amsterdam and the Netherlands

year	1970	1974	1975	1976	1977	1978	1979	1980	1981	1982
Amsterdam[1]			56,971		55,974		57,138		60,700	
Netherlands[2]	306,000	356,000		368,000		411,000		474,000		474,000

Source: 1) Bureau of Statistics of Amsterdam
                (<u>Amsterdam in Cijfers</u>, several volumes)
            2) EIM (1986), pp.19

An even more in-depth look however, - i.e., by considering those subsectors within the FIPS sectors which are most closely related to the more recent information 'technology system' - reveals that the Amsterdam region can again be considered as a predessor of developments taking place elsewhere. In this context, Alders and de Ruijter (1984), for example, have demonstrated that - as far as the more innovative parts of the nowadays highly dynamic producer service sector is concerned - in 1984 17% of these establishments with 22% of the total employment had their location in the Greater Amsterdam region.[7] In this context, Koerhuis and Cnossen (1982) estimate that as far as the new and innovative computer service and software firms is concerned, in 1981 19.4% of these establishments with 20.8% of total employment had their location in this region.

As indicated before, in Chapter 9 producer service sectors linked to the more recent 'information technology system' will be analyzed more in depth. In this chapter also the relative performance of the Amsterdam region compared to other (metropolitan) regions will be considered.

5.4 CONCLUSION

In this chapter structural economic developments within the Amsterdam region have briefly been considered. First of all, development patterns concerning the manufacturing sector have been analyzed. As stated in the first part of this book, in recent decades the employment generating capacity of this sector - resulting, inter alia, from a relative lack of basic innovations and related new technological 'trajectories' - has declined rather drastically (see for a further illustration of the Netherlands also Chapter 9 of this book).

As far as the *spatial* dimension is concerned however, it has been demonstrated that in general the Amsterdam region clearly *preceded* the industrial decline for the Netherlands as a whole. This pattern appears to support our theoretical framework in as far as the first hypothesis is concerned that the above-mentioned structural decline in later life cycles of industries and technologies will first be noticed in central (metropolitan) regions.

Also in more recent periods the Amsterdam region appears to perform relatively poor as far as its share in total manufacturing employment is concerned. This latter conclusion also holds for several other 'metropolitan' or central regions in the Netherlands. Consequently, these spatial patterns appear largely consistent with the 'urban-rural' shift of manufacturing employment observed for several EEC-countries (cf., Keeble et al 1983).

In the context of our theoretical framework, it is an interesting research issue to find out whether these spatial employment patterns are also reflected by - as stressed in our theoretical framework - spatial innovation trends. This issue will be taken up in Chapters 6 to 8. The empirical results presented in Chapter 7, for example, suggest that also in terms of the innovation potential of (small) manufacturing firms the Amsterdam region performs relatively poorly nowadays. Our results presented in that chapter suggest that this conclusion applies to both 'new line' and 'old line' sectors, be it especially for the latter types of sectors (see Figure 5.5 below).

As far as the 'take-off' of (relatively) new sectors - related to the more recent 'information technology system' (see Chapter 9) - is concerned however, the Amsterdam region again appears to be a predecessor of developments taking place elsewhere in the Netherlands. In this respect, it has been demonstrated here that in recent decades the employment base of this region has become increasingly biased toward finance, insurance and producer service activities.

In this context, the Amsterdam region also performs relatively favourably as regards its share in total employment in a sector largely generated by the more recent 'information technology system', i.e., computer services. Consequently, the Amsterdam region can be considered as a forerunner in a structural transformation process towards a post-industrial society (cf., Naisbitt 1984). In Chapter 9 it will be further demonstrated that the Amsterdam region can indeed be considered as an important incubation area of various (new and innovative) producer service activities.

Consequently, the broad structural transformations of the Amsterdam region described in this chapter - and our more in-depth empirical findings concerning this region to be presented in the next

chapters - indicate the following pattern as regards the relative
(innovative) performance of the Amsterdam region (see Figure 5.5
below).

Figure 5.5 Relative Performance of the Amsterdam Region

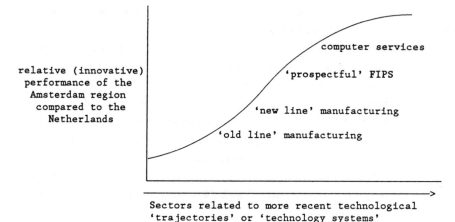

relative (innovative)
performance of the
Amsterdam region
compared to the
Netherlands

computer services

'prospectful' FIPS

'new line' manufacturing

'old line' manufacturing

Sectors related to more recent technological
'trajectories' or 'technology systems'

In this respect, the patterns described in the present chapter
also suggests that in the course of time structural developments in the
Amsterdam region seems to move from the top right hand side of this
figure to the left. In Chapter 9 this trend will be further illustrated
for various types of producer service sectors. This implies that in
order to keep its economic 'sustainability' the Amsterdam region is
'under constant pressure' to transform its economic base prior to
developments taking place elsewhere in the Netherlands.

Of course, it is difficult to predict the next stage of
developments in Amsterdam. Given its specific nodal position in an
international network, it seem plausible that international advanced
producer services (e.g., based on telecommunication, conferences,
consultancy activities, etc.) as well as advanced cultural and
educational facilities may be some of the core activities in future
developments of Amsterdam.

Consequently, the above patterns appear to be largely consistent
with our theoretical framework developed in the first part of this
book. More in particular the above patterns seem to confirm the first
hypothesis deduced from this framework in which it has been stated that

metropolitan areas like Amsterdam will be leading in the processes of growth and decline of (life cycles of) industries and technologies.

It is clear that a structural transformation incorporates a decline in some of the original base activities and an increase in new forms of product (or service) activities. Such structural changes have far reaching implications for the functioning of the labour market, but it seems appropriate to interpret such changes more positively than is usually done. In line with a Schumpeterian viewpoint, one might even claim that the search for new 'combinations' is a necessary condition for long-run sustainability of a city system.

Notes to Chapter 5:

1. 'Urban sustainability' is a key concept in the European network on Urban Innovation (URBINNO), sponsored by Volkswagen Foundation, in which the long-term evolution of various cities in Europe is taken into consideration. The present chapter is a result of a participation in this network.

2. This regional classification is based on an amalgamation of several so-called COROP-regions (see also the next chapter). For a definition of the classification of COROP-regions into central, intermediate and peripheral zone, the reader is referred to Chapter 7 of this book. In this context, Amsterdam, The Hague and Rijnmond in Table 5.4 refer to COROP regions with codes 231, 260 and 291, respectively.

3. As far as the COROP-region Utrecht is concerned - which is also largely 'metropolitan' in nature - the regional share in total manufacturing employment decreased from 5.1% in 1973 to 4.7% in 1986.

4. In 1973 4.2% of total employment in the manufacturing sector was located in the Arnhem-Nijmegen region, whilst in 1986 this figure amounted to 3.8%.

5. These data limitations also relate to the fact that as far as 'historical' data is concerned data are only available for the FIPS-sector as a whole (see also Table 5.6).

6. This location quotient is defined as a region's share in FIPS employment divided by its share in total employment.

7. This refers to the COROP region with code 230 (see for an explanation the next chapters).

6 Regional variations in innovative performance

6.1 INTRODUCTION

In this chapter our main attention will focus on measuring the impact of both *intra-firm* and *regional* variables on the innovative performance of different types of manufacturing firms - i.e., establishments - in the Netherlands.[1] As stated in Chapter 2, *spatial* variations in the innovative performance of firms may result from both a *non-homogeneous* spatial dispersion of firms having a high innovation potential - i.e., on the basis of their intra-firm characteristics- and an (additive) impact of location in a favourable production environment. In Chapter 2 the first mentioned impact has been labelled the '*structural*' component, while the second impact has been denoted as the '*production milieu*' component. Consequently, in this chapter the *first basic research issue*, already identified in Chapter 2, on the impact of both the 'structural' and 'production milieu' component on spatial variations in the innovative performance of firms will be analyzed.

In Chapter 2 it has been discussed that *empirical tests* of both components *simultaneously* are very rare in spatial innovation analysis. Exactly this empirical validation - i.e., of both impacts - will be the purpose of this chapter. So, besides the influence of (regional variations in) the intra-firm characteristics *also* the impact of location in a favourable production environment - ceteris paribus the intra-firm characteristics - on regional variations in innovative performance will be analyzed.

Given the more recently observed 'urban-rural' shift of manufacturing employment - i.e., from 'high quality production environments' to more 'remote (low quality) areas' (cf., Norton and Rees 1979, Bluestone and Harrison 1982, Keeble et al 1983, and for the Netherlands Wever 1985; see also the previous chapter) - it is an interesting research issue to find out whether also with respect to *innovative activities* these 'remote' areas are in a favourable position. This might (also) provide a (partial) explanation for this spatial shift of manufacturing employment. Also in our conceptual-theoretical framework developed in Chapter 3 we have called attention for the often ignored role of (further) innovative activities in such a spatial deconcentration trend.

In order to determine the impact of both components mentioned above - viz., the 'structural' and 'production milieu' component - it is necessary to have a reliable measuring rod of both the *innovation potential* and *innovativeness* of individual firms. As discussed in Chapter 1 however, so far a uniformly accepted definition has not been found. Consequently, various innovation indicators are currently being used, related to both the *input* - i.e., innovation potential - aspect (e.g., number of skilled workers, number of R&D employees) and the *output* - i.e., innovativeness - aspect (e.g., number of patents granted, number of new products) of the intra-firm innovative activities. However, it is increasingly realized that all such indicators provide at best a *partial* proxy for the complex nature of innovation processes. In this respect Hansen (1986, pp.2) remarks: 'As a result, the optimal approach has been the collection of a variety of types of information, none of which actually *measures* innovation, but all of which provide *indications* of the level of innovation'.

Also concerning the (supposed) impact of the 'production milieu' component a multiplicity of regional variables can be gathered from the literature. Several of these variables have been considered in Chapter 2 of this study (see in particular subsection 2.2.3).

Since various indicators are being used to approximate the variables *innovation potential* and *innovativeness* of individual firms as well as the 'quality' of the regional *'production milieu'*, it is more appropriate to regard these variables as *latent variables* which can be approximated by a set of observable - i.e., manifest-indicators. Such a *multivariate* approach brings us into the realm of latent variables models. Consequently, in the framework of our analysis we will use the *Partial Least Squares* (PLS) method (see Wold 1982, 1985a, 1985b) to determine the impact of both the 'structural' and 'production milieu' component on regional variations in the innovativeness of different types of (manufacturing) firms.

This method will briefly be considered in the next section. Then in section 6.3 we will discuss the various *indicators* and the (micro-level) database used to approximate the (intra-firm) latent variables *innovation potential* and *innovativeness*, respectively. Next, section 6.4 will be devoted to a discussion of the *regional indicators* used to approximate the latent variable 'production milieu'. Section 6.5 will describe the conceptual PLS model which will serve as a frame of reference in the following sections, while in section 6.6 our research methodology will be considered. In sections 6.7 - 6.9 the various empirical results will be presented while section 6.10 summarizes the most important conclusions.

6.2 THE PARTIAL LEAST SQUARES METHOD

As mentioned in the introduction to this chapter, usually a variety of (input and output) innovation indicators is used in innovation research. As a consequence, a multivariate analysis posesses important advantages (cf., Jöreskog and Wold 1982, Folmer 1985). In this regard, Fornell (1982, pp.19) states: "As is readily observed Lisrel and PLS are the most general and flexible methods (of multivariate analysis)". Both PLS and Lisrel can be considered as path models with latent variables that are indirectly observed by multiple manifest variables (MVs), called indicators (cf., Wold, 1985a).

The PLS method is obviously a *least squares oriented* approach. In this method no assumptions concerning the distributional properties of the variables have to be incorporated. In this respect Wold (1985a, pp.588) remarks: "in PLS modeling both estimation and evaluation are distribution - and independence - free".

The PLS method also generates explicit *case values*[2] for the latent variables. Even when PLS estimates are inconsistent, they are consistent at large, as "they tend to the true values when there is indefinite increase not only in the number of observed cases (N), but also in the number of indicators for each LV" (Wold 1985b, pp.231). In this context, our 'weakest' model (to be explained later on) has nearly 200 cases while every latent variable distinguished has several indicators. By means of simulation experiments Areskoug (1982), for example, has shown that for such a size the estimates of a two-latent variables PLS model are close to the 'true' values. Furthermore, "experience shows that if the arrow scheme is realistic in the sense that both PLS and Lisrel give small residuals, there is little or no difference between the PLS and Lisrel parameter estimates" (Wold 1985b, pp.241).

The Lisrel method is a *maximum likelihood* (ML) method which assumes the manifest (or measurable) variables to be jointly governed by a specified multivariate distribution which is subject to independent observation. Especially in the context of a regional analysis it may be questionable whether this ML assumption is realistic. Consequently, Apel (1980, pp.259) remarks: "with regard to the missing knowledge in most of the subjects of complex regional analysis, a statistically less pretentious method is called for".

Apart from the possible impairment of the 'hard' distributional ML assumptions, our analysis will also be 'compound' and consist of *many indicators*. In this context, it has been observed elsewhere that "the technical difficulty of Lisrel modeling increases rapidly with the size of the model" (Wold 1985a, pp.589) and that "the specification of the model can be seriously hampered by identification problems" (Apel 1980, pp.260).

Finally, in contrast to PLS, the Lisrel method does *not* generate explicit *case values* for the *latent variables* under consideration. As will be shown in the next chapter however, these case values provide very valuable information for the purposes of our analysis. On these grounds we considered the PLS method as an appropriate 'vehicle' for our analysis.

In PLS modelling - like in Lisrel - the analysis often starts by means of the construction of a conceptual arrow scheme - i.e., path model - indicating the relationships - and often the expected signs- between the latent variables (LVs) and their observable indicators or manifest variables (MVs). Thus in case of two LVs the following illustrative conceptual model might be specified (where according to convention squares represent MVs and circles LVs):

Figure 6.1 A Two Latent Variable Path Model

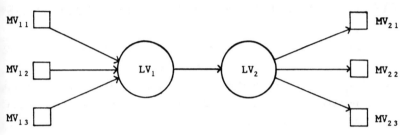

For details concerning all technicalities of the PLS estimation process the reader is referred to Wold (1982, 1985a, 1985b) and Lohmöller (1984). In short, the PLS estimation procedure aims at estimating the various parameter values between LVs mutually and between LVs and their manifest (observable) indicators. In this respect *three stages* can be distinguished:

- In the *first iterative* stage the *case values* of the *LVs* are generated. In this respect, each latent variable is estimated as a weighted aggregate of its indicators (manifest variables), with weights that are auxiliary parameters (cf., Wold 1985a). For each latent variable one has to choice between two estimation modes (mode A and B). For the choice of these modes, we refer to Lohmöller (1984) among others.

- In the *second stage* these estimated case values of the latent variables form the input for OLS regression procedures in order to estimate the parameter values for the relationships between the (estimated) LVs mutually - i.e., the *inner relations* - and

between these LVs and their indicators - i.e., the correlations
or *loadings* between each estimated LV and each belonging MV.
- In the *third stage* the *location parameters* can be derived. This
procedure is only useful when the measurement scales of the
observable indicators are comparable (cf., Lohmöller 1984).

After this concise discussion of PLS latent variables path
models, in the next section we will concentrate on the MVs used in our
analysis to approximate the intra-firm latent variables discussed in
the introduction to this chapter.

6.3 MEASURING INNOVATION POTENTIAL AND INNOVATIVENESS

As discussed in the previous sections, it is generally accepted
that both innovation potential and innovativeness of individual firms
can be considered as multi-dimensional compound phenomena. Consequently,
measuring innovative activities at the individual firm level is often
troublesome. In this respect, one of the essential problems appears to
be that the input and output indicators used (or available) are only
partly representative of these phenomena. As a consequence, the
explanatory power of models linking *one* single output indicator (for
example, number of product innovations per firm) to several input
indicators (e.g., number of R&D employees, total number of employees,
expenditures on external R&D) may be very poor (see for an illustration
Davelaar and Nijkamp 1987b, 1988).

In this respect, the LV approach of PLS, in which *multiple*
indicators are taken together in approximating both the input *and*
output aspect of innovative activities at the individual firm level,
may potentially be very promising. Consequently, in the present section
we will start with the *intra-firm* aspect of such an innovation model,
while the *regional* dimension will be the subject of the next section.

As a point of departure, in this section we will consider two
intra-firm concepts both of which will be regarded as LVs - i.e., the
innovation potential and *innovativeness* of an individual
(manufacturing) firm. As stated before, the first concept refers to the
input side of the innovation process, while innovativeness refers to
the *output* aspect of this process. We expect both concepts to be

positively related.

So in our analysis the LV *innovation potential* refers to the endowment of a firm - i.e., a manufacturing establishment (see below)- with innovation generating or stimulating factors. In this respect, several of these input variables have been formulated in the literature. All these variables are expected to exert a *positive* influence on the capacity to generate (or adopt) innovations at the individual firm level.

In our framework several of these indicators will be combined in a 'master concept' called *innovation potential*. Besides the evident advantage of using multiple indicators in determining the innovation potential of a firm, such an approach may also be helpful in assessing the *relative importance* of the several *input indicators* distinguished. In this respect, the question whether the importance of the selected input indicators varies among *different types of sectors* will also be considered.

Secondly, in relation to the *output* side of the innovation process, *innovativeness* will be considered as a LV which is reflected in *several* (partial) *output indicators*. So also innovativeness will be conceived of as a multi-dimensional phenomenon which needs to be measured by means of multiple ouput indicators. Consequently, we will select several of these indicators all of which are expected to be representative for the LV *innovativeness*. Also in this case an important advantage appears to be that the relative importance of the successive output indicators - for various types of sectors - can be gauged (see also below).

Because both *innovation potential* and *innovativeness* are determined by *multiple indicators*, the estimated relations between these LVs might be much more indicative of the strength of this input-output relation than the linkage of one single output indicator with several input indicators (as in a regression analysis, for example) would suggest.

We will now discuss the micro-level database - and the MVs (i.e., indicators) - used in our following PLS analysis. Although- also from a technical point of view - it would be attractive to use a large series of indicators, data limitations restrict this. Therefore, we have used the results of an extensive postal inquiry by Kleinknecht among nearly 3,000 main *establishments* (larger than 10 employees) - to be *labelled firms* in this study - of the *manufacturing sector* in the Netherlands. The *response rate* of this inquiry was *63.1%* (i.e., 1,842 firms). The questions in this inquiry (see also below) referred to the innovative activities performed within the firms in *1983* (for details concerning this inquiry, see Kleinknecht 1987b).

From these disaggregate data several indicators - related to both the LVs innovation potential and innovativeness - have been derived. After several experiments we selected the following indicators[3] from this inquiry:

A. INNOVATION POTENTIAL (INPUT) INDICATORS:

a) **R** : Number of *employees* involved in *R&D* activities within the firm. In this context definitions of R&D activities were based on the Frascati-manual as developed by the OECD.

b) **EXTRD** : Number of *employees* involved in *external R&D* - e.g., in universities, technical institutes, other firms - employed on behalf of the firm.

c) **RDO** : Actual *growth of R&D efforts* during the past three years. This indicator has been specified as a dummy (because of the measurement scale of this variable) which will be set equal to 1 in case of *positive* growth. This indicator tries to incorporate some of the *dynamic* aspects of the *innovative input efforts*. This is also the purpose of the next indicator.

d) **IN** : *Expected growth of R&D* efforts in the next two years (1984 and 1985). This indicator has also been specified as a dummy which will be set equal to 1 in case the firm expected a *positive* growth.

e) **W** : Total *number of employees* of the firm. This indicator tries to capture the scale at which the firm is operating.

f) **OMZ$_1$** : *Development of sales* compared to 1982. Because of the measurement scale this variable has also been specified as a dummy which will be set equal to 1 in case sales expanded more than 10%. Clearly this indicator is meant to reflect the past growth of the firm activities.

g) **OMZ$_2$** : *Expected growth of sales* in the next year (1984). This dummy will be set equal to 1 in case it is expected that sales will increase by *more than 10%*. This indicator is meant to reflect the future growth perspectives of the firm.

h) **EXPORT:** *Export orientation* of the firm. This dummy will be set equal to 1 if the firm exported more than 25% of its sales in 1983. Especially in a small country such as the Netherlands this indicator is often thought to be positively related to the innovation potential of the firm (cf., Kleinknecht and Verspagen 1987, Tuyl 1987).

The indicators a) - h) make up our set of manifest variables used to approximate the intra-firm LV *innovation potential*. Our a priori expectations are that all above mentioned indicators will be *positively* related to this LV.

B. INNOVATIVENESS (OUTPUT) INDICATORS:

a) **I$_1$** : Number of product innovations (new to the firm) introduced *during 1983* (this holds for all the ouput indicators mentioned below).

b) **I$_2$** : Number of process innovations (new to the firm).

c) **I$_3$** : Number of 'combined' - i.e., both product and process - innovations (new to the firm).[4]

d) **RI$_1$** : Number of product innovations (i.e., a subset of a)) which were - according to the firm - new to the whole branch of industry in which the firm operates.

e) **RI$_2$** : The same with respect to process innovations.

f) **RI$_3$** : The same with respect to the 'combined' innovations.

As stated before, it may sometimes be difficult for firms to interpret the definitions of innovations given in the inquiry and, consequently, to assess the exact *number* of innovations. In this regard, one could imagine two 'extreme' possibilities for firms, viz., those exhibiting a 'low' threshold value with respect to labeling something an 'innovation' and those having 'high' threshold values.

Clearly, the first group will report more innovations than the second group. However, when we would hold a survey after the *average preparation time* of the reported innovations, the answers of the first group would in general tend to be lower (as they also include 'low quality' innovations). Consequently, we have also compiled the following 3 (multiplicative) output indicators:

g) VI_1 : *Average* preparation time[5] (in months) of the product innovations mentioned under a) * I_1 = *Total* preparation time (in months) of the product innovations new to the firm introduced during 1983.

h) VI_2 : The same with respect to process innovations new to the firm.

i) VI_3 : The same with respect to the 'combined' innovations new to the firm.

We expect the intra-firm LV innovativeness (again) to be *positively* related to all indicators mentioned above.

After this discussion of the intra-firm dimension of the innovation process, the *regional* dimension will be considered in the next section.

6.4 INDICATORS OF THE REGIONAL 'PRODUCTION MILIEU'

As discussed in Chapter 2, the literature concerning the identification of (supposed) regional stimuli for the (intra-firm) innovation process is extensive. It may suffice to indicate here that a great many regional variables have been proposed, several of which have been discussed in Chapter 2. Empirical tests concerning the relevance of these variables is scarcer however, due to the following causes:

1) Coherent *regional data* sets concerning these (supposed) regional stimuli are generally *scarce* and often only available at a few (rather aggregate) spatial levels.

2) There is a serious *lack of micro (firm based) data* sets by means of which the impact of these regional stimuli - ceteris paribus the intra-firm characteristics - can be assessed. Consequently, one is often forced to use very small data sets or to use rather

indirect or crude proxies (e.g., number of patents granted in the region, number of firms in the regions that are thought to have favourable technological prospects). It needs no comment that in this way the determination of the (additive) impact of these regional stimuli on the innovativeness of individual firms is seriously being hampered.

As our micro level data discussed in the previous chapter refer to the Netherlands, we will now confine our discussion of the regional dimension of innovative activities to the Dutch context. Even in a small country like the Netherlands, the spatial (economic) variations may - depending on the subject of study - sometimes be quite considerable. In this respect, it is common to subdivide the Netherlands into 40 standard statistical (nodal) regions - called *'COROP' regions* - which from a socio-economic point of view can be considered to be rather *homogeneous*. This geographical subdivision is the most important regional scale at which regularly geographical socio-economic data are provided by, for example, the Central Bureau of Statistics.

In the context of *innovative activities*, the Netherlands Economic Institute (1984) has recently constructed several regional indicators with respect to each COROP region. One of the intentions of this study was to gain a more reliable (quantitative) insight into the quality of the regional 'production milieu' - especially with respect to innovative activities - prevailing in these 40 COROP regions.

Consequently, our set of regional indicators used to approximate the LV 'production milieu' will largely be based on the several indicators constructed by the Netherlands Economic Institute supplemented by a few indicators collected by us.

The regional *'production milieu'* indicators - constructed for all COROP regions separately - which will be used in the following analysis are the following.[6]

C. PRODUCTION MILIEU INDICATORS:

a) **BEVDCO** : Density of population.

b) **NEIOPCO** : Average educational level of the workforce.

c) **NEIZNCO** : Distance - in minutes of travelling time - to the national (economic) centre of gravity.

d) **NEIZICO** : Distance to the international (European) economic centre of gravity.

e) **NEIKCO** : Availability of office buildings.

f) **NEIBTCO** : Availability of building sites.

g) **NEIBGCO** : Availability of industrial buildings.

h) **KEAOPP** : Number of establishments that are considered to have favourable technological opportunities[7] - according to a selection made by the Netherlands Economic Institute (1984) - divided by the total number of manufacturing establishments.

i) **KEAWOP** : Total number of employees working in these selected (see indicator h) above) establishments divided by the total number of manufacturing establishments.

j) **NEIDUM1** : Availability of knowledge centres.

k) **NEIDUM3** : Accessibility via waterways.

l) **NEIDUM5** : Availability of communication infrastructure.

m) **NEIDUM6** : Availability of multiple enterprise buildings (with respect to innovative activities).

n) **NEIDUM7** : Agglomeration economies.

o) **NEIDUM9** : Quality of living environment.

p) **NEIDUM11**: Favourable institutional and policy framework.

The regional indicators have been specified in such a way that a higher regional score implies a 'better' production milieu, the only exception being the 'distance' indicators NEIZNCO and NEIZICO where a lower score implies closer proximity to the (inter)national economic centre.

Indicators b) and d) - g) have been measured as indices, while the (qualitative) indicators j) - p) have been specified as dummies which have been set equal to 1 in case a COROP-region performs relatively well for the indicator considered (for details, see appendix

1 of this chapter).

In order to incorporate the urban dimension and to test whether intra-COROP differences are important - e.g., between between urban and non-urban areas within a COROP region - we also introduced the following additional dummy variables:

q) **DUM1**   : This dummy variable has been set equal to 1 in case a firm (in the inquiry) is located in a so-called C4 or C5 municipality - according to a classification of Dutch municipalities by the Central Bureau of Statistics - i.e., a municipality with a central city larger than 50,000 inhabitants. In this study these municipalities will be denoted as the urban areas (within a COROP region).

r) **DUM2**   : This dummy variable equals 1 in case a firm is located in a B3 municipality - i.e., a municipality in which more than 30% of the (male) labour force commutes. As will be clear, these municipalities are often suburban.

6.5 THE CONCEPTUAL PLS MODEL

In section 6.3 we discussed several intra-firm input and output indicators for the innovation process, while in the previous section various regional indicators were introduced. These indicators will subsequently be integrated in a conceptual 'full' PLS model that will be estimated in the next sections (see Figure 6.2 below).

Figure 6.2 The Conceptual 'Full' PLS Model

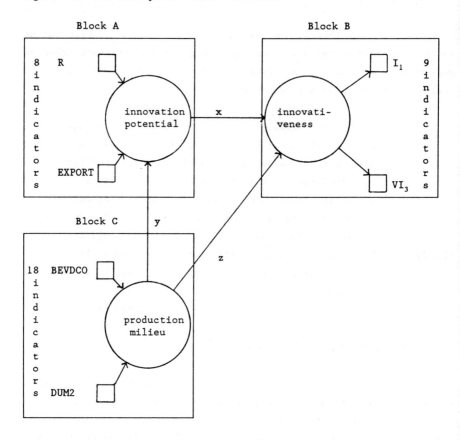

It should be recalled from the previous sections that the indicators of blocks A and B have been measured at the micro (i.e., firm) level, while the indicators of block C have been measured at the regional level.[8]

Because regional variations in the innovative performance - i.e., innovativeness - of individual firms might be due to a non-homogeneous spatial dispersion of firms having a high innovation potential[9] and/or an (additive) 'production milieu' impact we linked the LV 'production milieu' to both the (intra-firm) LVs innovation potential and innovativeness. Consequently, by means of the first link - i.e., inner relation y in Figure 6.2 above - it can be checked whether regions are equally well endowed with 'innovation potential' firms (i.e., the 'structural' component), while the second link - i.e., inner relation z - tries to determine the (additive) *'production milieu'* impact.

Since the LVs have been measured by means of several indicators, the estimated *inner* relations between the LVs mutually may give a more reliable picture of the strength of the interrelations between these concepts than the use of single indicators would reveal.

On the other hand, the estimated *outer relations* - i.e., the relationship between the LV and its indicators - of block A provide information concerning the relative importance of the various intra-firm (input) indicators in shaping the innovation potential of a firm. This also holds true for the indicators of block C with respect to the LV 'production milieu'. Regarding the indicators of block B, the results of the estimation procedure can be helpful in deciding on the question which innovation output indicators are most closely related to - i.e., provide the best reflection of - the (intra-firm) LV innovativeness. Thus a whole series of interesting research question can now be dealt with. In the next sections we will basically concentrate on the following selected issues:

1) For the inner relations: to what extent are the LVs innovation potential, innovativeness and 'production milieu' interrelated? In this respect, we expect a *positive* relation for the link between the (intra-firm) LVs *innovation potential* and *innovativeness*. For the relationship between the LVs *'production milieu'* and *innovation potential no a priori expectation* will be formulated with respect to the sign of parameter y. Depending on the locational profile of firms with a high innovation potential, this parameter could either be positive or negative. Concerning the pair of LVs *production milieu-innovativeness* parameter z could either be *close to zero* - i.e., rejecting the hypothesis of an additive (positive) production milieu impact - or (strongly) *positive* - i.e., providing empirical evidence of such an impact.

2) For the *outer* relations: which indicators are important in shaping (block A and C) or reflecting (block B) the three LVs mentioned above?

As the 'answers' to both issues mentioned above may vary according to types of firms considered, we will perform our analysis for *four types of firms* in the inquiry:

a) *Small* firms (W ≤ 100) operating in 'old line' or 'traditional' sectors like food, textile, wood processing (see also Kok et al 1985 and Stokman 1985). In this group (small) firms with (2-digit) SIC-codes 20-27 have been included. This group will be labelled the 'old line' firms.

b) The second group will consist of *small* (W ≤ 100) firms having (2-digit) SIC-codes 28-39 (for example, chemical, metal, electronics). This group will be labelled the 'new line' firms.

c) The third group distinguished will consist of those *small* (W ≤ 100) firms with (4-digit) SIC-codes which - according to the selection made by the Netherlands Economic Institute (see the previous section) - are expected to have rather favourable (technological) prospects. This group is a (smaller) subset of the type b) firms mentioned above. These firms will be denoted as small 'technologically promising' firms.

d) The fourth group distinguished will consist of *all* firms in the sample having (4-digit) SIC-codes that belong to the *group of selected SIC-codes* by the Netherlands Economic Institute. So, besides the small firms mentioned under c), in this group the larger (W > 100) firms will also be included.

In terms of the first part of this book, it can be stated that in general sectors (firms) belonging to *group a* are linked to more *former* technological *'trajectories'* than the *type b* sectors (firms). On the other hand, the *type c* sectors (firms) are a *selected subset* of all 'new line' (i.e., group b) sectors which are again linked to *more recent 'trajectories'* than the *type b* sectors as a whole. So, generally speaking, if we move from type a to type b and c (or d) firms, we move from sectors linked to former 'trajectories' to sectors linked to more recent 'trajectories'.

In our analysis we will mainly concentrate on small firms, with the exception of type d) firms. The reasons for this are the following:

1) Large firms often produce of multiplicity of products and therefore their SIC classification may sometimes be more or less arbitrary.
2) The inclusion - or analysis - of large firms often implies a high variance with respect to some indicators of (especially) block A (R, EXTRD,W) and block B.
3) Innovations may be more comparable between small firms mutually than between small and large firms.
4) Small firms are often expected to be more sensitive to regional impacts than large firms (cf., Giaoutzi et al, 1988).

To determine whether - in spite of the above-mentioned remarks - the main conclusions derived with respect to type a) till c) firms would change when large firms are included we will - as an example - also analyze the type d) firms distinguished above.

The empirical results concerning the types of firms mentioned above will be presented in sections 6.7 till 6.9. First of all however, in the next section our research methodology will briefly be considered.

6.6 THE RESEARCH METHODOLOGY

For each type of firm distinguished in the foregoing section, the 'full' PLS model (cf., Figure 6.2) has been estimated.[10] As an example the estimated inner relations and loadings - i.e., the single correlation between the estimated LVs and their indicators (see before) - with respect to the *type a* firms have been presented in Table 6.1 below.[11]

Table 6.1   Results of the 'Full' PLS Model (estimated inner relations
and loadings) with respect to Type a Firms

Inner relation		Innovation Potential		Innovativeness		Production Milieu			
coeff.	est.	ind.	load.	ind.	load.	ind.	load.	ind.	load.
x	0.58	R	0.53	I_1	0.68	KEAOPP	0.36	NEIDUM1	0.40
y	-0.24	EXTRD	0.55	I_3	0.50	KEAWOP	0.05	NEIDUM3	0.31
z	-0.01	EXPORT	0.48	I_2	0.54	BEVDCO	0.34	NEIDUM5	0.34
		RDO	0.60	RI_1	0.64	NEIOPCO	0.68	NEIDUM6	-0.07
		IN	0.61	RI_3	0.37	NEIZNCO	-0.39	NEIDUM7	0.29
		W	0.50	RI_2	0.53	NEIZICO	-0.06	NEIDUM9	0.38
		OMZ1	0.12	VI_1	0.69	NEIKCO	0.31	NEIDUM11	-0.28
N = 442		OMZ_2	0.37	VI_3	0.46	NEIBTCO	-0.20	DUM1	0.24
R^2 = 0.34				VI_2	0.54	NEIBGCO	0.18	DUM2	-0.15

In interpreting the results of the several tables which will be
presented in this and the next chapter, it should be kept in mind that
both the *indicators* and the *LVs* have been standardized to *unit variance
and mean zero* in order to be able to compare the relevance of
individual indicators as well as the inner relations between the LVs.

Before interpreting these results in more detail, two important
conclusions can already be drawn from Table 6.1 above:

1) In contrast to blocks A and B, the loadings of the indicators in
block C display a *diffuse pattern* with most indicators being
positively and a few negatively related to the LV 'production
milieu'.

2) Some, especially regional, indicators appear to be only very *weakly
related* to their *LV* (notably 'production milieu').

In order to circumvent the *first* mentioned problem, our
subsequent research strategy has been to 'divide up' the LV 'production
milieu' of the 'full' PLS model into a regional *latent 'pull' variable*
- called *REGPULL* and a regional *latent 'push' variable* - called
REGPUSH.[12] The regional indicators of the LV *REGPULL* will consist of
those regional indicators which appeared to have *negative* loadings with
respect to the LV *'production milieu'* of the 'full' PLS model.[13] On the

other hand, the regional indicators of the LV *REGPUSH* will consist of those regional indicators which were *positively* related to the LV '*production milieu*' of the 'full' PLS model.

As regards the *second issue* raised above, we (successively) removed those regional indicators having low loadings - i.e., in general those with absolute values smaller than 0.20 with the exception of DUM1 and DUM2 - with respect to their LVs REGPULL or REGPULL.

Consequently, our 'restricted' conceptual PLS model then becomes the following (see Figure 6.3, where only the latent variables are sketched).

Figure 6.3   The 'Restricted' PLS Model

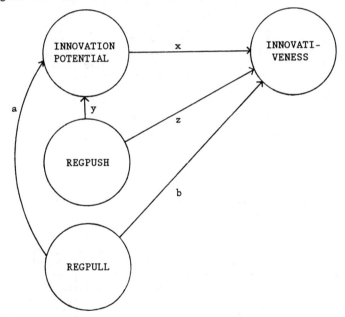

We shall present the 'final' results of our analysis. In the next section we will start with the small 'old line' firms.

6.7 PLS RESULTS FOR THE SMALL 'OLD LINE' FIRMS

The final results of the *restricted* PLS model concerning the *type a* firms distinguished in section 6.5 can be found in Table 6.2 below. For the purposes of our analysis it is interesting to analyze the relative performance of the *three large metropolitan areas* in the Netherlands - i.e., Amsterdam, Rijnmond and The Hague. Therefore, we also estimated a 'restricted' PLS model net of these areas.[14] The final results concerning this model version can be found in Table 6.3 below.

Table 6.2 PLS Estimates for Small 'Old Line' Firms

Inner relation		Innovation Potential		Innovativeness		Production Milieu REGPUSH		REGPULL	
coeff.	est.	ind.	load.	ind.	load.	ind.	load.	ind.	load.
x	0.58	R	0.54	I_1	0.68	BEVDCO	0.42	NEIZNCO	0.83
y	-0.20	EXTRD	0.53	I_3	0.50	KEAOPP	0.44	NEIBTCO	0.46
z	0	EXPORT	0.49	I_2	0.54	NEIOPCO	0.81	NEIDUM11	0.59
a	0.06	RDO	0.60	RI_1	0.64	NEIKCO	0.38	DUM2	0.36
b	0.03	IN	0.60	RI_3	0.37	NEIDUM1	0.47		
		W	0.49	RI_2	0.54	NEIDUM3	0.38		
		OMZ1	0.12	VI_1	0.69	NEIDUM5	0.41		
		OMZ_2	0.38	VI_3	0.46	NEIDUM7	0.35		
N = 442				VI_2	0.55	NEIDUM9	0.46		
R^2 = 0.34						DUM1	0.29		

Table 6.3 PLS Estimates for Small 'Old Line' Firms (minus three metropolitan areas)

Inner relation		Innovation Potential		Innovativeness		Production Milieu REGPUSH		REGPULL	
coeff.	est.	ind.	load.	ind.	load.	ind.	load.	ind.	load.
x	0.58	R	0.57	I_1	0.67	KEAOPP	0.48	NEIZNCO	0.78
y	-0.17	EXTRD	0.53	I_3	0.54	NEIOPCO	0.83	NEIBTCO	0.57
z	-0.02	EXPORT	0.49	I_2	0.54	NEIKCO	0.30	NEIDUM11	0.47
a	0.07	RDO	0.61	RI_1	0.65	NEIDUM1	0.43	DUM2	0.43
b	0.05	IN	0.61	RI_3	0.42	NEIDUM3	0.43		
		W	0.45	RI_2	0.55	NEIDUM7	0.31		
		OMZ1	0.13	VI_1	0.69	NEIDUM9	0.41		
		OMZ_2	0.40	VI_3	0.50	DUM1	0.18		
N = 401				VI_2	0.54				
R^2 = 0.34									

In these tables N refers to the number of cases while R^2 is the squared multiple correlation of the endogenous *LV innovativeness* with respect to the other (explanatory) LVs.[15]

From these tables various interesting conclusions can be drawn with respect to the type a firms. Here we will mainly concentrate on some of the most interesting issues. On the basis of Tables 6.2 and 6.3 the following conclusion can be drawn.

The estimated inner relation between the (intra-firm) LVs *innovation potential* and *innovativeness* is - as a priori expected- clearly *positive*. In this respect, the most important indicators for the *innovation potential* of these firms are the 'dynamic' indicators 'past growth of internal R&D' (RDO) and 'expected growth' of this variable (IN). This might be an indication of the fact that for these firms it is important to make a distinction between 'stagnating' and 'growing' - i.e., in terms of R&D efforts - firms. Also the indicators representing the level of R&D efforts (R, EXTRD) appear to be important in shaping the innovation potential of these firms. On the other hand, (especially) development of sales in the recent *past* (OMZ_1) and expectation concerning *future* sales (OMZ_2) appear to be less important in this context.

With regard to *block B*, the indicators representing *product innovations* (I_1, RI_1, VI_1) appear to be the *best* representative indicators of *innovativeness*. On the other hand, the 'combination' indicators (I_3, RI_3, VI_3) in general appear to be the weakest representations of the innovativeness of small 'old line' firms. It is interesting to observe here however, that also *all* indicators representing process innovations (i.e., I_2, RI_2 and VI_2) have rather *high loadings* - compared to the other types of firms distinguished (see the next sections) - with respect to the LV innovativeness. Apparently, the generation (or adoption) of process innovations is a more important aim of the innovative efforts of small 'old line' firms than for the other types of firms. This is in accordance with the 'innovation life cycle' concept discussed in the first part of this study. According to this concept the generation (or adoption) of process innovations will become increasingly important in later phases of the life cycles of industries and technologies. In Chapter 8 of this study, the issue of product versus process innovation will be further analyzed.

As to the *spatial* dimension, it appears that the estimated inner relations referring to *regional variations* in the (intra-firm) *innovation potential* (parameters y and a of Figure 6.3) - i.e., the 'structural' component - are far more important than the inner relations reflecting the (additive) 'production milieu' impact. As can be derived from both Tables 6.2 and 6.3 above, the estimated inner relations representing the latter impact - i.e., parameters z and b in Figure 6.3 - are very small and contrary to the à priori hypothesis of a *positive* impact (in the case of coefficient z in Table 6.3).

Consequently, regional variations in the *innovativeness* of type a firms in first instance appears to be due to a *non-homogenenous* spatial dispersion of firms having a (relatively) high or low *innovation potential* on the basis of their intra-firm characteristics and *not* to an additional - i.e., after compensating for this non-homogeneous dispersion - 'production milieu' impact as such.

In this respect, it is interesting to observe that several of the selected regional indicators - often considered to be conducive to a 'high quality' production milieu - appear to be *positively* related to the LV *REGPUSH* which in turn is quite *negatively* related (cf., parameter y) to the (intra-firm) LV *innovation potential*. Consequently, regions having favourable scores on these indicators - i.e., in general the COROP regions located in the Rimcity - are apparantly not well endowed with type a firms having a favourable (intra-firm) innovation potential.

This is also suggested by the fact that four selected regional indicators appear to be *positively* linked to the LV *REGPULL* which by itself is (weakly) *positively* related to both *innovation potential* and *innovativeness*. It is noteworthy, that especially concerning these indicators COROP regions located in the central (western) part of the Netherlands perform relatively poorly (see appendix 1).[16] In the next chapter, this issue will be analyzed more in-depth.

In relation to the intra-regional (i.e., intra-COROP) dimension of innovation, it is worth noting that the urban areas within a COROP region (for which the DUM1 indicator has been set equal to 1) do not appear to attract the high innovation potential firms of type a either. This is reflected by the fact that the *DUM1* indicator is *positively* related to the LV *REGPUSH* which by itself is *negatively* related to the LV *innovation potential*. On the other hand, 'commuter municipalities' (represented by DUM2) appear to perform more favourable given the fact that the *DUM2* indicator is *positively* related to the LV *REGPULL* (which is positively related to the LV innovation potential).

In case the three large metropolitan areas are removed from the sample (see Table 6.3) the positive loading of the DUM1 indicator with respect to the LV REGPUSH decreases. Apparently, firms located in these metropolitan areas are less well endowed with innovation potential than firms located in the other urban areas. In the next chapter, the (regional) issues raised above will be further analyzed. In that chapter the relative performance of the three metropolitan areas separately will also be considered. First of all however, in the next section attention will be focused on the small 'new line' firms.

6.8 PLS RESULTS FOR THE SMALL 'NEW LINE' FIRMS

In this section the results of our 'restricted' PLS model concerning type b firms will be presented. Table 6.4 below summarizes our estimations for these firms. After 'elimination' of type b firms in the inquiry which are located in one of the three metropolitan areas mentioned before, we find the PLS estimations which are included in Table 6.5 below.

Table 6.4 PLS Results for Small 'New Line' (type b) Firms

Inner relation		Innovation Potential		Innovativeness		Production Milieu REGPUSH		REGPULL	
coeff.	est.	ind.	load.	ind.	load.	ind.	load.	ind.	load.
x	0.56	R	0.70	I_1	0.72	BEVDCO	0.54	KEAWOP	0.30
y	-0.09	EXTRD	0.55	I_3	0.53	NEIOPCO	0.27	NEIZNCO	0.30
z	0	EXPORT	0.62	I_2	0.54	NEIZICO	0.47	NEIBTCO	0.58
a	0.06	RDO	0.72	RI_1	0.58	NEIKCO	0.39	NEIDUM11	0.68
b	0.04	IN	0.51	RI_3	0.35	NEIDUM1	0.30		
		W	0.43	RI_2	0.28	NEIDUM3	0.78		
		OMZ1	0.31	VI_1	0.65	NEIDUM5	0.47		
		OMZ2	0.30	VI_3	0.53	NEIDUM7	0.60		
N = 716				VI_2	0.48	NEIDUM9	0.51		
R^2 = 0.32						DUM1	0.24		
						DUM2	0.11		

Table 6.5 PLS Results for Small 'New Line' Firms (minus three metropolitan areas)

Inner relation		Innovation Potential		Innovativeness		Production Milieu REGPUSH		REGPULL	
coeff.	est.	ind.	load.	ind.	load.	ind.	load.	ind.	load.
x	0.60	R	0.64	I_1	0.68	NEIOPCO	0.41	KEAWOP	0.33
y	-0.05	EXTRD	0.67	I_3	0.57	NEIDUM1	0.47	KEAOPP	0.35
z	-0.03	EXPORT	0.55	I_2	0.55	NEIDUM3	0.44	NEIBTCO	0.58
a	0.10	RDO	0.67	RI_1	0.55	NEIDUM6	0.46	NEIDUM11	0.48
b	0.02	IN	0.50	RI_3	0.36	NEIDUM7	0.44	DUM1	0.27
		W	0.43	RI_2	0.29	NEIDUM9	0.63		
		OMZ1	0.26	VI_1	0.61	DUM2	0.27		
		OMZ2	0.27	VI_3	0.57				
N = 638				VI_2	0.47				
R^2 = 0.36									

On the basis of these tables the following conclusions can be drawn.

The intra-firm LVs *innovation potential* and *innovativeness* are again clearly *positively* related. As regards these LVs, especially the level of internal R&D efforts (R) and past development of these efforts (RDO) appear to be important indicators of the *innovation potential* of small 'new line' firms. In this respect, the loading of the indicator reflecting internal R&D efforts is larger than in the case of small 'old line' firms. Apparently, these R&D efforts are more important in determining the innovation potential of small 'new line' firms than for the type a firms analyzed in the previous section. Next in importance, as regards the LV innovation potential, appear to be the indicators reflecting export orientation (EXPORT), external R&D (EXTRD) and (future) development of R&D efforts (IN).

Just like the type a firms discussed in the previous section, indicators reflecting the *product* oriented innovative activities of small 'new line' firms (I_1, RI_1, VI_1) are the best representative indicators of the intra-firm LV *innovativeness*. Rather remarkable in this case however, is the low loading of the *RI2 indicator* as compared to the small 'old line' firms. The generation (or adoption) of entirely new - i.e., new to the whole branch of industry in which the firm operates - *process innovations* is apparently not one of the major aims of the innovative efforts of small 'new line' firms. In Chapter 8 of this study we will return to this issue.

As far as the *regional dimension* is concerned, also in this case no clear empirical *evidence* of an (additive) '*production milieu*' impact can be gathered from these tables. Also for the 'new line' firms the estimated inner relations (z and b in Figure 6.3) reflecting this impact are low in absolute value and - in the case of Table 6.5 - have opposite signs. Again - given the larger (absolute) values of the estimated inner relations y and a (see Figure 6.3) - the '*structural*' *component* appears to be *more important* in determining spatial variations in the innovative performance of type b firms.

Also in this case, several regional indicators reflecting a 'high quality' production environment - see Table 6.4 - are *positively* related to the LV REGPUSH which is *negatively* related (cf., the estimated inner relation y) - to the intra-firm LV *innovation potential*. Thus (also) concerning the small 'new line' firms there appears to be a discrepancy between the spatial dispersion of firms having a high (intra-firm) innovation potential and the 'quality' of the regional production environment.

In this respect, especially the COROP regions located in the central western parts of the Netherlands - i.e., the Rimcity - have rather favourable scores on those *indicators* which are *positively* related to the LV *REGPUSH*. On the other hand, these COROP regions (again) perform less well concerning the regional indicators which are *positively* related the the LV *REGPULL* (which is positively related to the intra-firm LV innovation potential).[17] So, again this suggests that these centrally located COROP regions are rather poorly endowed with small 'new line' firms having a relatively high innovation potential. In the next chapter, this issue will be further analyzed.

As regards the (intra-regional) urban dimension, in Table 6.4 the *DUM1* indicator appears to be *positively* related to the LV *REGPUSH*. After removing those firms which are located in the three large metropolitan areas however, this indicator becomes positively related to the LV *REGPULL* (which has a positive estimated inner relation with respect to the LV innovation potential). So, also concerning the small 'new line' firms this indicates a rather poor performance of these metropolitan areas (compared to the other urban areas for which the DUM1 indicator equals 1). The 'commuter municipalities', on the other hand, now appear to perform less well than in the case of small 'old line' firms. This is reflected by the fact that the *DUM2* indicator is now *positively* related to the *LV REGPUSH*. These intra-regional issues will also be investigated further in the next chapter.

In the next section our estimation results with respect to type c and d firms will be presented.

6.9 PLS RESULTS FOR THE TECHNOLOGICALLY PROMISING FIRMS

In this section we will concentrate on type c and d firms distinguished in section 6.5. The results of the 'restricted' PLS concerning type c firms are summarized in Table 6.6 below. In Table 6.7 firms located in the three large metropolitan areas are again excluded.

Table 6.6 PLS Results for Type c Firms

| Inner relation | | Innovation Potential | | Innovativeness | | Production Milieu | | | |
| | | | | | | REGPUSH | | REGPULL | |
coeff.	est.	ind.	load.	ind.	load.	ind.	load.	ind.	load.
x	0.58	R	0.73	I_1	0.62	BEVDCO	0.50	NEIBTCO	0.50
y	-0.23	EXTRD	0.18	I_3	0.63	NEIOPCO	0.42	NEIDUM11	0.71
z	-0.04	EXPORT	0.62	I_2	0.56	NEIZICO	0.50	DUM1	0.52
a	0	RDO	0.62	RI_1	0.49	NEIDUM1	0.50		
b	0.04	IN	0.44	RI_3	0.53	NEIDUM3	0.50		
		W	0.33	RI_2	0.26	NEIDUM5	0.60		
		OMZ1	0.36	VI_1	0.57	NEIDUM7	0.54		
		OMZ2	0.09	VI_3	0.60	NEIDUM9	0.52		
N = 221				VI_2	0.54	DUM2	0.27		
R^2 = 0.35									

Table 6.7 PLS Results for Type c Firms (minus 3 metropolitan areas)

| Inner relation | | Innovation Potential | | Innovativeness | | Production Milieu | | | |
| | | | | | | REGPUSH | | REGPULL | |
coeff.	est.	ind.	load.	ind.	load.	ind.	load.	ind.	load.
x	0.59	R	0.76	I_1	0.54	NEIOPCO	0.36	KEAOPP	0.22
y	-0.26	EXTRD	0.35	I_3	0.66	NEIZICO	0.39	NEIKCO	0.18
z	-0.05	EXPORT	0.58	I_2	0.57	NEIDUM1	0.38	DUM1	0.94
a	0.14	RDO	0.57	RI_1	0.47	NEIDUM3	0.30		
b	0.03	IN	0.44	RI_3	0.55	NEIDUM5	0.51		
		W	0.34	RI_2	0.29	NEIDUM6	0.30		
		OMZ1	0.42	VI_1	0.50	NEIDUM7	0.28		
		OMZ2	0.10	VI_3	0.63	NEIDUM9	0.39		
N = 183				VI_2	0.51				
R^2 = 0.37									

Some of the most important conclusions that can be drawn from these tables are the following. Like before, the estimated inner relation between the intra-firm LVs *innovation potential* and *innovativeness* is again clearly *positive*. In this respect, internal R&D (R), (past) development of internal R&D efforts (RDO) and export orientation (EXPORT) are - in this rank order - the three most important indicators concerning the LV *innovation potential* of the firms under consideration. Quite remarkable in this case however, is the rather low loading of the indicator reflecting *external* R&D efforts compared to the other types of firms discussed in the previous sections. On the other hand, the indicator reflecting *internal* R&D efforts (R) appears to be more important - given the larger loading with respect to this indicator - than for type a and b firms analyzed in the previous sections. As regards the LV innovation potential, expectations concerning future sales (OMZ_2) appear to be quite irrelevant. Consequently, firms with optimistic expectations do not seem to be more inclined to increase their generation (or adoption) of innovations in the present.

All indicators of block B reflect the LV *innovativeness* rather well with the only *exception of RI_2* - i.e., process innovations new to the whole branch of industry in which the firm operates. As discussed in the previous sections, a similar conclusion could be drawn for the small 'new line' firms, but *not* for the small 'old line' firms. Consequently, the generation (or adoption) of relatively new - i.e., new to the whole branch of industry - *process* innovations does *not appear* to be one of the *major aims* of the innovative activities of these type c firms. As stated before, this pattern is in agreement with the 'innovation life cycle' concept discussed in the first part of this study.

Also concerning type c firms our results do *not point at* a positive impact of the '*production milieu*' component on the (intra-firm) level of *innovativeness*. In both tables, the estimated inner relations representing this impact - i.e., parameters z and b - are again low in absolute value and contrary in sign. In this respect, the '*structural*' component (cf., inner relations a and y) - pointing at a *non-homogeneous spatial dispersion* of type c firms having a relatively high or low *innovation potential* on the basis of their intra-firm characteristics - again appears to be (far) more important.

Also in this case however, the more innovation potential firms of type c do not appear to be located in the 'high quality' production environments. Several of the selected regional indicators - reflecting these 'high quality' environments - are (again) *positively* related to the LV *REGPUSH* which is *negatively* related to the intra-firm LV *innovation potential*. In the next chapter we will return to this issue.

In comparing the results of Tables 6.6 and 6.7 our results (again) indicate that the 'performance' of the three metropolitan areas mentioned before is rather extra-ordinary. According to Table 6.6 the LV *REGPULL* is *unrelated* to the the intra-firm LV *innovation potential*. After removing the (type c) firms located in these metropolitan areas however, the role of the LV *REGPULL* - i.e., concerning its positve impact on the LV *innovation potential* - increases considerably while also the *DUM1 indicator* becomes very *important* in determining this regional LV.

From these results it can be concluded that there are - according to Table 6.7 - clear intra-regional (i.e., intra-COROP) differences with respect to the innovation potential of type c firms. At the intra-regional level, type c firms located in the 'urban' (DUM1) areas in general appear to be better endowed with innovative capacity than firms located in the 'non-urban' areas (see also the next chapter). As discussed above however (see the results with respect to Table 6.6), this conclusion does not hold for the type c firms located in the three large metropolitan areas.

In case *all* firms in the sample which are considered to have rather favourable technological prospects - i.e., on the basis of their SIC-codes as selected by the Netherlands Economic Institute (1984)- are included, we find the results for type d firms in Table 6.8 below. Like before, in Table 6.9 type d firms located in one of the three metropolitan areas have been excluded.

Table 6.8 PLS Estimates with respect to Type d Firms

| Inner relation | | Innovation Potential | | Innovativeness | | Production Milieu | | | |
| | | | | | | REGPUSH | | REGPULL | |
coeff.	est.	ind.	load.	ind.	load.	ind.	load.	ind.	load.
x	0.66	R	0.90	I_1	0.66	BEVDCO	0.20	NEIZNCO	0.31
y	-0.24	EXTRD	0.65	I_3	0.57	KEAOPP	0.21	NEIDUM11	0.41
z	-0.07	EXPORT	0.44	I_2	0.34	NEIOPCO	0.17	DUM1	0.81
a	0.07	RDO	0.42	RI_1	0.68	NEIZICO	0.42		
b	0	IN	0.30	RI_3	0.52	NEIBGCO	0.34		
		W	0.55	RI_2	0.13	NEIKCO	0.33		
		OMZ1	0.27	VI_1	0.82	NEIDUM1	0.17		
		OMZ_2	0.01	VI_3	0.74	NEIDUM3	0.40		
				VI_2	0.44	NEIDUM5	0.28		
						NEIDUM6	0.19		
N	= 323					NEIDUM7	0.20		
R^2	= 0.46					NEIDUM9	0.27		

Table 6.9   PLS Estimates with respect to Type d Firms (minus 3 metropolitan areas)

| Inner relation | | Innovation Potential | | Innovativeness | | Production Milieu | | | |
| | | | | | | REGPUSH | | REGPULL | |
coeff.	est.	ind.	load.	ind.	load.	ind.	load.	ind.	load.
x	0.69	R	0.87	I_1	0.69	NEIDUM1	0.37	KEAWOP	0.23
y	-0.23	EXTRD	0.82	I_3	0.56	NEIOPCO	0.33	DUM1	0.99
z	0.01	EXPORT	0.43	I_2	0.37	NEIZICO	0.46		
a	0.09	RDO	0.40	RI_1	0.71	NEIDUM3	0.35		
b	0.06	IN	0.33	RI_3	0.49	NEIDUM5	0.46		
		W	0.64	RI_2	0.14	NEIBGCO	0.39		
		OMZ1	0.25	VI_1	0.83	NEIDUM9	0.45		
		OMZ_2	0.02	VI_3	0.74	DUM2	0.06		
N = 268				VI_2	0.44				
R^2 = 0.48									

In the context of our analysis only the most important conclusions, which can derived from Tables 6.8 and 6.9 above, will briefly be considered here. As found for the other types of firms, the estimated inner relation between the intra-firm LVs *innovation potential* and *innovativeness* is again clearly *positive*. As regards these intra-firm LVs, the indicators reflecting *internal R&D* (R) and *external R&D* efforts (EXTRD) appear to be the most important variables in shaping the *innovation potential* of type d firms. Especially *internal R&D* appears to be important in this respect (also compared to the results for the other types of firms discussed in the previous sections). Concerning the LV innovation potential, export orientation (EXPORT, (past) development of internal R&D efforts (RDO) and number of employees (W) constitute the second group of important indicators. This indicator appeared to be less relevant when analyzing small firms (see Tables 6.2 till 6.7). The inclusion of large firms raises the importance of this indicator. As regards this intra-firm LV, expectations concerning future sales (OMZ_2) appear to be wholly irrelevant.

With regard to block B the LV *innovativeness* is very weakly reflected in the generation (or adoption) of *process* innovations new to the entire branch of industry (RI_2). In general, *all indicators* reflecting the *process* oriented innovative activities of these firms (I_2, RI_2, VI2) are relatively *weakly related* to this *LV*. Apparently, the (major) aims of the innovative efforts of type d firms are not oriented towards process innovations (see also type c firms). In this respect, the indicators representing *product* innovations (I_1, RI_1, VI_1) are far more important. The indicators having the highest loadings with respect to the LV *innovativeness* however, appear to be our own 'constructs' VI_1 and VI_3. This may be due to the fact that here we are studying all size categories of firms and consequently, these indicators become important yardsticks to compare the innovativeness of small and large firms mutually.

The results concerning the *regional* dimension of innovative activities appear to be largely consistent with our prior analysis. Also with respect to type d firms no empirical evidence of a '*production milieu*' impact on the innovative performance of these firms could be gathered from the data. The estimated inner relations representing this impact are low and/or contrary to the à priori expectation of a *positive* impact (i.e., in the case of parameter z in Table 6.8) resulting from location in a 'high quality' production environment.

Like before, the non-homogeneous spatial dispersion of type d firms having a high or low (intra-firm) potential to innovate - i.e., the '*structural*' component - appears to be (far) more important in determining spatial variations in innovative performance (see the estimated inner relations a and y). Also in this case, several regional indicators reflecting 'high quality' production environments are *positively* related to the LV *REGPUSH* which is quite strongly *negatively* related to the LV *innovation potential*. Consequently also with respect to *type d* firms regions offering 'high quality' environmental conditions in general do *not* appear to possess the more innovation potential firms.

As in the case of type c firms discussed above, the *DUM1* indicator is (strongly) *positively* related to the LV *REGPULL* which is *positively* related to the intra-firm LV *innovation potential*. This indicates the same type of *intra-regional* (intra-COROP) *differences* - between urban and non-urban areas within a specific COROP - discussed in the context of type c firms (i.e., after excluding the three metropolitan areas, see Table 6.7). Contrary to type c firms discussed above however, these intra-regional differences also apply when the three metropolitan areas are included. However, this *pattern* becomes *more pronounced* in Table 6.9 where these areas have been excluded. From

this table it can be derived that *both* the loading of the *DUM1*
indicator with respect to the LV *REGPULL* and the estimated (positive)
inner relation between this LV and the intra-firm LV *innovation
potential* have *increased*.

6.10 CONCLUSION

In this chapter the impact of spatial and non-spatial variables
on the *innovation potential* and *innovativeness* of individual
manufacturing firms (i.e., establishments) has been analyzed. In this
respect, the *first basic research issue* identified in Chapter 2 of this
study, on the impact of both the *'structural'* and *'production milieu'*
component on spatial variations in the innovative performance of firms
has been analyzed.

In this context, the *'structural'* component refers to the *non-
homogeneous* spatial dispersion of firms having a high (or low)
innovation potential on the basis of their *intra-firm characteristics*.
The *'production milieu'* component on the other hand, refers to the
(additive) impact - ceteris paribus the intra-firm characteristics (or
'structural' component) - of location in a *'high quality'* production
environment on the innovative performance of these firms.

To this purpose, both the *innovation potential* and *innovativeness*
of (individual) firms have been conceived as (multi-dimensional) *latent
variables*. In using an extensive survey among manufacturing
establishments in the Netherlands, we selected several indicators-
i.e., manifest variables - to approximate these intra-firm LVs. The
indicators related to the LV *innovation potential* are notably input
variables of the innovation process such as internal and external R&D
efforts. On the other hand, the LV *innovativeness* has been approximated
using output indicators such as the number of product and process
innovations. As to the *spatial* dimension, we selected several regional
indicators often thought to reflect 'high quality' production
environments (for details see section 6.4).

In this context, *four types of firms* have been distinguished. This
typology was based on the SIC-codes of the firms in the sample. In terms

of the first part of this study, the different types of firms
distinguished - ranging from 'old line' to 'technologically promising'
firms - can (in general) be considered to be related to technological
'trajectories' of different time periods.

As to our empirical results, the importance of the several
(intra-firm) indicators related to the LVs *innovation potential* and
innovativeness varied according to *type of firm* analyzed. Concerning the
LV *innovativeness* one may, for example, refer to the fact that
indicators reflecting *process* innovations (I_2, RI_2, VI_2) did not perform
very well in the analysis of firms that are expected to have rather
favourable technological prospects - i.e., type c and d firms - in
contrast with the small 'old line' firms. With respect to the LV
innovation potential, for example, the importance of the indicator
reflecting internal R&D efforts (R) appeared to be more important for
type c and d firms mentioned above than for the small 'old line' firms
(for more specific results, see the foregoing sections).

As to the *regional* dimension, *spatial variations* in the endowment
of (COROP) regions with firms having a relatively high (or low) intra-
firm *innovation potential* - i.e., the 'structural' component discussed
above - appeared to be fairly important. On the other hand however, *no
clear empirical evidence* of an (additive) impact of the *'production
milieu'* component on the intra-firm level of *innovativeness* could be
gathered from the data. For all types of firms distinguished in this
chapter, the estimated inner relations reflecting this impact appeared
to be low in absolute value and/or contrary to the a priori expectation
of a positive effect resulting from location in a 'high quality'
production environment.
Consequently, the 'explanation' of regionally varying degrees of
innovativeness (of manufacturing firms) appears in the first place to be
due to a *non-homogeneous spatial dispersion* of firms having a high (or
low) *innovation potential* on the basis of their intra-firm
characteristics. After 'compensating' for this 'structural' component
(cf., the inner relations a and y in Figure 6.3) a (positive) regional
'production milieu' impact per se could not be deduced from the data.

With respect to the *'structural'* component discussed above, an interesting research finding appeared to be the fact that - for all types of firms analyzed - several of the selected *regional indicators* reflecting 'high quality' production environments were (strongly) *positively* related to the regional *LV REGPUSH*. On the other hand, this latter LV was quite *negatively* related to the LV *innovation potential*. This suggests that those regions that are generally considered to provide the most *favourable* ('high quality') *production environments*-i.e., in general the COROP regions located in the Rimcity - are rather *poorly endowed* with firms having a relatively high *innovation potential*. In the next chapter, the spatial dispersion of firms having a relatively high or low (intra-firm) innovation potential will be further analyzed.

Notes to Chapter 6:

1. In the following pages the terms 'firms' and 'establishments' will be used interchangeably, both referring to manufacturing establishments.

2. Case values are defined here as weighted aggregates of the constituent indicators of a latent variable (see, for example, Wold 1982; see also below).

3. As stated above, these indicators refer to 1983 unless stated otherwise.

4. Product innovations were defined as *new* - or (technically) improved-*products*. Process innovations were defined as *new* production *process* technologies concerning *existing* products, while the 'combined' innovations were defined as both new (or technically improved) products produced by new process technologies.

5. In the inquiry firms were only asked for the average preparation time of the innovations mentioned under a) till c).

6. For an overview of the 'performance' of each COROP region with respect to the selected indicators: see appendix 1 of this chapter.

7. These establishments have been selected on the basis of their 4-digit SIC-classification. This selection is largely based on a 'translation' of the (technological) promising *activities* for the Dutch economy - as identified by the so-called Wagner-commission - to *SIC-codes*. In this context, some service sectors (like computer-service) have been included as well. As regards the manufacturing sector a rather broad group of 37 (4-digit) sub-sectors was selected. Consequently, the selected sub-sectors are not denoted as 'high tech' by the Netherlands Economic Institute. In this respect, the group of sub-sectors selected is also far more extensive than, for example, the (10) 'high-tech' sub-sectors selected by Bouwman et al (1985a, 1985b) for the Dutch economy.

8. In our analysis we used PLS mode B in estimating the weight relationships of the blocks A and C, while PLS mode A has been applied with respect to block B (for more details concerning the choice of weight modes, the reader is referred to Wold 1982 and Lohmöller 1984).

9. This refers to the 'structural' component identified in Chapter 2 and the introduction to this chapter.

10. Because of a disproportionate influence on the estimation caused by some implausible answers (e.g., small firms without R&D claiming to have large numbers of innovations or very large firms of type d with many employees in R&D claiming to have no innovations at all) about 2 per cent of the most extreme cases have been deleted a priori from our sample.

11. As regards the estimated 'weights', the covariance between the indicators may sometimes hamper the interpretation of these 'weights'. For example, a high positive covariance between two indicators of one LV may result in a positive 'weight' for the one indicator and a negative 'weight' for the other indicator while the single correlations

- i.e., loadings - between the estimated LV and both indicators are clearly positive. In this respect, 'the loading p_h measures the separate contribution of X_h to the relevance of its LV. The weight w_h measures the contribution of indicator X_h to the joint relevance of the indicators' (Wold 1982, p.18).

12. The terms 'pull' and 'push' refer to the relationship of these regional LVs with the intra-firm LV innovation potential.

13. In the case of type a) firms - see Table 6.1 - these indicators are: NEIZNCO, NEIZICO, NEIBTCO, NEIDUM6, NEIDUM11 and DUM2.

14. In this model version, firms located in Amsterdam (i.e., sub-COROP region 231), Rijnmond (i.e., sub-COROP region 291) and the Metropolitan Area The Hague (i.e., COROP region 260) have been excluded from our sample.

15. Results of Tukey's jack-knife tests for each final 'restricted' PLS model with respect to the parameters of the endogenous LV innovativeness can be found in appendix 2 of this chapter (for details concerning these tests, see Lohmöller 1984 among others).

16. Recall that a higher score on NEIZNCO implies a larger distance to the national economic centre of gravity.

17. For details, see appendix 1 of this chapter.

Appendix 1 'Performance' of regions with respect to selected regional indicators

INDICATOR COROP REGION	(a) BEVDCO 1-1-'83	(b) NEIOPCO	(c) NEIZNCO	(d) NEIZICO	(e) NEIKCO	(f) NEIBTCO	(g) NEIBGCO	(h)	(i)
10. Oost-Groningen	189	76	173	200	15	188	28	86	937
20. Delfzijl	230	94	190	208	55	200	137	34	603
30. Ov. Groningen	274	110	163	196	50	200	68	247	4,388
40. Nrd. Friesland	203	91	135	198	84	162	53	222	2,386
50. Zuidw. Friesl.	141	91	121	193	73	200	0	78	665
60. Zuido. Friesl.	168	87	147	195	5	182	26	128	1,544
70. Nrd. Drenthe	165	115	148	190	9	165	132	77	1,691
80. Zuido. Drenthe	184	80	150	192	19	200	28	86	2,518
90. Zuidw. Drenthe	136	80	121	178	12	160	85	69	802
100. Nrd. Overijsse	178	93	96	172	48	138	78	241	2,732
110. Zuidw. Overij.	312	99	92	163	63	66	93	78	1,563
120. Twente	393	92	133	180	67	55	151	426	13,868
130. Veluwe	307	94	84	161	27	45	98	491	6,104
140. Achterhoek	234	93	93	163	19	49	105	279	2,735
150. Arnhem Nijmeg.	661	102	84	146	39	78	82	403	9,097
160. Zuidw. Geld.	260	77	60	160	13	111	128	156	1,267
170. Utrecht	693	109	34	163	126	31	127	724	13,129
180. Kop v. Nrd.Hol.	299	106	95	207	43	160	68	232	2,091
190. Alkmaar e.o.	653	115	55	189	109	97	141	146	998
200. IJmond	1117	122	43	184	0	15	22	101	1,596
210. Aggl. Haarlem	1628	123	37	182	37	43	110	188	3,425
220. Zaanstreek	1475	100	42	184	84	55	167	113	2,109
230. Gr. Amsterdam	4104 (595)	111	44	185	177	89	129	1,112	12,755
240. Gooi en Vecht	1303	116	48	169	71	3	87	308	6,962
250. Leiden en Bol.	1419	109	22	187	154	11	171	191	1,458
260. Agg. Den Haag	3095	111	27	182	204	12	53	576	9,590

INDICATOR COROP REGION	(j)	(k)	(l)	(m)	(n)	(o)	(p)	(s) total no. of man. est. 1-1-1984
10. Oost-Groningen	0	0	0	0	--	--	+	392
20. Delfzijl	0	++	0	0	--	--	+	109
30. Ov. Groningen	++	+	0	+	0	0	+	998
40. Nrd. Friesland	0	+	0	0	-	0	+	836
50. Zuidw. Friesl.	0	+	0	0	--	0	+	491
60. Zuido. Friesl.	0	0	0	0	--	-	+	533
70. Nrd. Drenthe	0	0	0	0	-	0	+	305
80. Zuido. Drenthe	0	0	0	0	--	-	+	355
90. Zuidw. Drenthe	+	0	0	0	--	0	+	345
100. Nrd. Overijsse	0	0	0	0	-	0	+	1,067
110. Zuidw. Overij.	0	0	0	0	-	0	+	344
120. Twente	++	0	0	+	0	0	+	1,954
130. Veluwe	+++	0	0	0	0	++	+	1,714
140. Achterhoek	0	0	0	0	--	-	+	1,421
150. Arnhem Nijmeg.	++	+	0	0	+	+	+	1,743
160. Zuidw. Geld.	0	+	0	0	-	-	+	870
170. Utrecht	+++	+	0	0	+	+	0	2,902
180. Kop v. Nrd.Hol.	+	+	0	0	-	-	0	839
190. Alkmaar e.o.	0	0	0	0	+	0	0	592
200. IJmond	0	++	0	0	++	0	0	381
210. Aggl. Haarlem	0	+	0	0	++	++	0	690
220. Zaanstreek	0	+	0	0	++	-	0	636
230. Gr. Amsterdam	+++	++	+	0	++	+	0	4,299
240. Gooi en Vecht	0	0	0	0	++	++	0	968
250. Leiden en Bol.	++	0	0	0	++	0	0	829
260. Agg. Den Haag	++	+	0	0	++	+	0	1,807

INDICATOR COROP REGION	(a) BEVDCO 1-1-'83	(b) NEIOPCO	(c) NEIZNCO	(d) NEIZICO	(e) NEIKCO	(f) NEIBTCO	(g) NEIBGCO	(h)	(i)
270. Delft en Wstl.	1072	100	33	182	51	12	69	192	6,566
280. Oost Zuid Hol.	570	105	16	175	136	31	168	257	3,454
290. Gr. Rijnmond	1946 (263)	90	31	175	143	31	100	825	17,352
300. ZO Zuid-Holl.	724	95	39	169	94	57	126	336	10,618
310. Zeeuws-Vlaand.	148	93	142	195	37	200	69	68	3,026
320. Ov. Zeeland	234	94	115	190	31	200	29	170	4,906
330. West Nrd. Brab.	413	96	65	158	15	200	139	455	9,345
340. Midden Nrd. Br.	413	93	77	150	40	74	72	318	6,940
350. Noordo.Nrd. Br.	608 (303)	96	65	150	49	66	134	328	11,233
360. Zuido. Nrd. Br.	458	105	88	139	98	60	104	448	5,061
370. Nrd. Limburg	299	92	126	120	11	98	31	163	7,702
380. Mid. Limburg	316	95	125	121	29	165	26	167	2,665
390. Zuid. Limburg	916	87	148	111	27	98	57	440	15,378
400. ZIJP	107	122	85	181	651	58	41	93	869

Source: Source: NEI (1984)
Reg. Stat.
Zakboek,
pp.229

INDICATOR COROP REGION	(j)	(k)	(l)	(m)	(n)	(o)	(p)	(s) total no. of man. est. 1-1-1984
270. Delft en Wstl.	+++	0	0	0	++	0	0	676
280. Oost Zuid Hol.	0	0	0	0	+	+	0	1,143
290. Gr. Rijnmond	+	+++	0	0	++	0	0	3,549
300. ZO Zuid-Holl.	0	+	0	0	+	0	0	1,216
310. Zeeuws-Vlaand.	0	++	0	0	--	-	+	357
320. Ov. Zeeland	0	++	0	0	-	0	+	669
330. West Nrd. Brab.	0	+	0	0	0	0	+	1,778
340. Midden Nrd. Br.	0	+	0	0	0	0	+	1,832
350. Noordo.Nrd. Br.	0	+	0	0	0	0	+	1,685
360. Zuido. Nrd. Br.	++	0	0	+	+	+	+	2,284
370. Nrd. Limburg	0	+	0	0	-	0	+	980
380. Mid. Limburg	0	+	0	0	-	0	+	857
390. Zuid. Limburg	+	+	+	0	+	0	+	1,708
400. ZIJP	+	+	0	+	-	0	+	253

Source: NEI (1984)

Source: Regionaal
Statistisch Zakboek
1984, pp.282

Legend:

(a) With respect to COROP regions 230, 290 and 350 the first terms refer to the population densities of the metropolitan areas within these regions - viz., the sub COROP regions Amsterdam (231), Rijnmond (291) and Stadgewest Den Bosch (351) - while the densities of their immediate surroundings - i.e., sub COROP regions 232, 292 and 352 - are presented within brackets.
(h) Number of establishments - ultimo 1983 - with favourable technological prospects (on the basis of their SIC-codes). Indicator KEAOPP has been defined as column (h) divided by column (s).
(i) Number of employees working in these establishments (see (h)). Indicator KEAWOP has been defined as column (i) divided by column (s).
(j) Availability knowledge centres. NEIDUM1 = 1, in case of ++ or +++, NEIDUM1 = 0, otherwise.
(k) Accessibility via waterways. NEIDUM3 = 1, in case of ++ or +++, NEIDUM3 = 0, otherwise.
(l) Communication infrastructure. NEIDUM5 = 1, in case of + NEIDUM5 = 0, otherwise.
(m) Multiple enterprise buildings. NEIDUM6 = 1, in case of + NEIDUM6 = 0, otherwise.
(n) Agglomeration economies. NEIDUM7 = 1, in case of ++ NEIDUM7 = 0, otherwise.
(o) Quality living environment NEIDUM9 = 1, in case of + or ++ NEIDUM9 = 0, otherwise.
(p) Institut. and policy framework NEIDUM11= 1, in case of + NEIDUM11= 0, otherwise.

Appendix 2. Jack-knife results for the parameters of the
endogenous LV innovativeness.[18]

Table 1   Jack-knife results for type a firms

Inner relation			Loadings		
parameter	mean	standard deviation	indicator	mean	standard deviation
x	0.57	0.01	I_1	0.69	0.02
z	-0.001	0.006	I_3	0.51	0.03
b	0.03	0.01	I_2	0.55	0.02
			RI_1	0.64	0.03
			RI_3	0.37	0.02
			RI_2	0.54	0.04
			VI_1	0.70	0.03
			VI_3	0.47	0.02
			VI_2	0.56	0.02

Table 2   Jack-knife results for type a firms (minus three
metropolitan areas)

Inner relation			Loadings		
parameter	mean	standard deviation	indicator	mean	standard deviation
x	0.57	0.01	I_1	0.67	0.02
z	-0.02	0.01	I_3	0.55	0.03
b	0.04	0.007	I_2	0.55	0.02
			RI_1	0.65	0.03
			RI_3	0.42	0.04
			RI_2	0.55	0.06
			VI_1	0.70	0.03
			VI_3	0.51	0.03
			VI_2	0.55	0.03

Table 3   Jack-knife results for type b firms

Inner relation			Loadings		
parameter	mean	standard deviation	indicator	mean	standard deviation
x	0.54	0.008	I_1	0.72	0.02
z	0	0.03	I_3	0.54	0.04
b	0.04	0.01	I_2	0.54	0.03
			RI_1	0.56	0.03
			RI_3	0.34	0.02
			RI_2	0.28	0.02
			VI_1	0.66	0.03
			VI_3	0.54	0.05
			VI_2	0.49	0.03

[18] In this respect, the estimated means and standard errors are based on an omission
distance of 9 (for details, see Lohmöller 1984 among others).

Table 4   Jack-knife results for type b firms (minus three
metropolitan areas)

Inner relation			Loadings		
parameter	mean	standard deviation	indicator	mean	standard deviation
x	0.58	0.005	I_1	0.68	0.02
z	-0.03	0.01	I_3	0.57	0.03
b	0.02	0.01	I_2	0.55	0.03
			RI_1	0.55	0.02
			RI_3	0.36	0.03
			RI_2	0.30	0.02
			VI_1	0.61	0.01
			VI_3	0.58	0.03
			VI_2	0.47	0.02

Table 5   Jack-knife results for type c firms

Inner relation			Loadings		
parameter	mean	standard deviation	indicator	mean	standard deviation
x	0.57	0.01	I_1	0.63	0.05
z	-0.04	0.01	I_3	0.63	0.07
b	0.04	0.007	I_2	0.57	0.05
			RI_1	0.51	0.06
			RI_3	0.55	0.09
			RI_2	0.27	0.05
			VI_1	0.59	0.04
			VI_3	0.61	0.08
			VI_2	0.54	0.04

Table 6   Jack-knife results for type c firms (minus three
metropolitan areas)

Inner relation			Loadings		
parameter	mean	standard deviation	indicator	mean	standard deviation
x	0.58	0.02	I_1	0.55	0.07
z	-0.05	0.01	I_3	0.65	0.08
b	0.03	0.02	I_2	0.58	0.06
			RI_1	0.49	0.05
			RI_3	0.55	0.09
			RI_2	0.30	0.04
			VI_1	0.52	0.06
			VI_3	0.64	0.07
			VI_2	0.52	0.05

Table 7. Jack-knife results for type d firms

Inner relation			Loadings		
parameter	mean	standard deviation	indicator	mean	standard deviation
x	0.67	0.05	I_1	0.66	0.04
z	-0.068	0.01	I_3	0.58	0.03
b	0	0.01	I_2	0.33	0.08
			RI_1	0.69	0.04
			RI_3	0.53	0.05
			RI_2	0.12	0.07
			VI_1	0.81	0.08
			VI_3	0.74	0.06
			VI_2	0.43	0.08

Table 8. Jack-knife results for type d firms (minus three metropolitan areas)

Inner relation			Loadings		
parameter	mean	standard deviation	indicator	mean	standard deviation
x	0.68	0.06	I_1	0.68	0.05
z	0.006	0.01	I_3	0.56	0.04
b	0.06	0.02	I_2	0.37	0.10
			RI_1	0.70	0.05
			RI_3	0.48	0.05
			RI_2	0.14	0.06
			VI_1	0.81	0.08
			VI_3	0.73	0.07
			VI_2	0.44	0.09

7 Spatial variations in innovation potential

7.1 INTRODUCTION

In the foregoing chapter the regional dimension of innovative activities with respect to several types of Dutch manufacturing firms (i.e., establishments) has been analyzed. This analysis shows that there are (considerable) spatial variations in the endowment of regions with respect to firms having a relatively high (or low) intra-firm capacity to innovate, i.e., innovation potential. After compensating for this 'structural' component, a regional 'production milieu' impact - resulting from location in a 'high quality' production environment- on (spatial variations in) the innovativeness of individual firms could not be identified from the data.

Consequently, in this chapter we will concentrate on the former- i.e., 'structural' - component. In this context, we will consider the 'performance' of several (types of) regions - i.e., on the basis of the industrial survey discussed in the previous chapter - with respect to firms having a relatively high or low (intra-firm) potential to innovate. This analysis will be executed for the *various types of small* - i.e., type a, b and c firms - manufacturing firms distinguished in the previous chapter.[1] It goes without saying that the following results concerning the 'performance' of *individual* COROP regions - especially in the case of a low number of observations - should be interpreted carefully. In this context, the main purposes of the present chapter will be to *further illustrate* the results of the foregoing chapter as well as to indentify some *general trends* which are important for the purposes of our analysis.

As discussed in Chapters 1 and 5 of the present study (see also Chapter 9), the manufacturing sector rather rapidly lost its employment generating capacity in recent decades. Generally speaking, the lack of new technological 'trajectories' combined with a decline in the possibilities to generate further product innovations along previously established 'trajectories' had a negative impact upon manufacturing employment in these periods. In this respect, the generation of cost-efficient (labour-saving) process innovations became increasingly important (cf., Van Duijn 1981, 1983 Freeman et al 1982, Rothwell and Zegveld 1985; see also Chapter 1).

As to the *spatial* dimension, especially the more centrally located (metropolitan) regions appeared to have suffered from this decline in manufacturing employment (cf Keeble et al 1983, Wever 1985; see also Chapter 5). In the context of our analysis, it is an interesting research issue to find out whether this pattern also holds for the indigenous innovation potential of these regions. In our theoretical framework discussed in the first part of this study, we also called attention for the - theoretically often ignored - role of (further) innovative activities in such a spatial deconcentration trend in later phases of the life cycles of industries and technologies. In this respect, the third hypothesis deduced from this framework states that in such later phases non-metropolitan or non-central regions will become dominant concerning the (further) innovative activities performed.

Consequently, in order to test this hypothesis, in the present chapter (more general) spatial innovation patterns for the Dutch manufacturing sector will be analyzed. In this respect it will, for example, also be analyzed whether the relative poor performance of the Amsterdam area - as regards development of manufacturing employment in recent decades (see Chapter 5) - is also reflected in a weak innovation potential of (several types of) manufacturing firms. Thus, the analysis of the present chapter will provide a second empirical test of the validity of the theoretical framework developed in the first part of this study.

To these purposes, in the next section our research strategy concerning the selection of (COROP) regions will be discussed. In this context, we will discriminate between a 'positive' and a 'negative' set of regions. The former will in general consist of regions characterized by a relatively *favourable* score (i.e., case values) of its firms (in the sample) on the firm specific LV *innovation potential* discussed in the previous chapter. The opposite conclusion holds for the (COROP) regions which will be included in the 'negative' set of regions. In section 7.3 this research strategy will be applied to the small 'old line' firms. The small 'new line' and 'technologically promising' firms will be considered in sections 7.4 and 7.5, respectively. Finally, section 7.6 will provide some general conclusions. In this latter section, the empirical results of this and the previous chapter will also be briefly reconsidered in the light of our theoretical-conceptual framework discussed in the first part of this study.

7.2 THE SELECTION STRATEGY

In the foregoing chapter, two *regional* latent variables have been distinguished, viz., the LV *REGPULL* and the LV *REGPUSH*. From the analysis in Chapter 6 it appeared that the last mentioned LV - viz., the LV REGPUSH - was *negatively* related to the *intra-firm* LV *innovation potential*. The REGPULL variable on the other hand, was *positively* linked to this intra-firm LV.

In this context, various regional indicators have been used to approximate both these regional LVs. As stated in Chapter 6, most of these indicators - viz., the indicators a) - p) of section 6.4- referred to the *COROP level* of analysis. In order to determine whether also intra-regional (i.e., intra-COROP) differences in innovative activities were important however, two regional indicators referring to the *municipality* level - i.e., the DUM1 and DUM2 indicators - have been incorporated. We will indicate areas - i.e., municipalities - for which the *DUM1 indicators equals 1^2* as *'DUM1'* or *'urban'* areas. In a similar vein, areas for which the *DUM2 indicator equals 1* will be referred to as *'DUM2'* areas.

From the above arguments, it can be concluded that with respect to each regional LV discussed above *potentially* three - i.e., when both the DUM1 and DUM2 indicators belong to the set of regional indicators used to approximate the LV REGPULL or REGPULL - different areas within each COROP region can be distinguished.[3] Taking the example of small 'new line' firms, the results of Table 6.4 in section 6.8 of the previous chapter indicate that *both* the DUM1 and DUM2 indicator belong to the set of indicators used to approximate the LV REGPUSH.[4] Consequently, as regards these firms *three* different *types of areas*— each having a *specific case value for the LV REGPUSH* - can be distinguished within a specific *COROP region A*:

1. The *non-DUM1* and *non-DUM2* areas within COROP region A. As will be clear, the *case values* concerning the *LV REGPUSH* are the *same* for all (firms located in) these areas of COROP A. This also holds for the (firms located in) the type 2 and 3 areas of COROP region A:
2. The *'DUM2'*-areas within COROP region A.
3. The *'DUM1'*-areas within COROP region A.

As the DUM1 and DUM2 indicators are non-region (i.e., non-COROP) specific - i.e., they apply to the 'DUM1' and 'DUM2' areas of *all* COROP-regions concerned - the mutual discrepancies as regards the case values for the LV REGPUSH are the *same* for the type 1, 2 and 3 areas within each COROP region (see also the next section). Consequently, the case value for this LV is in *all* COROP regions influenced to the *same extent* by a *firm location* in a *DUM1* or *DUM2* area *vis-a-vis* location in a *type 1 area*.[5]

Concerning the *selection* of COROP regions, the following remarks are in order. We will discrimate between a *'positive'* and *'negative'* group of COROP regions. In general, the first group will contain those COROP regions which are relatively *well endowed* with (type a, b or c) firms having *favourable scores on the intra-firm LV innovation potential* (see Chapter 6), while the opposite conclusion holds for the 'negative' group.

In order to further illustrate the empirical results of the previous chapter, as a *first criterion* we will select those COROP regions having rather *'extreme'* case values for the regional *LVs REGPUSH and/or REGPULL* considered in the previous chapter. As will be clear from the empirical results found in Chapter 6, these regions will in general be characterized by rather 'extreme' (positive or negative) scores of its firms on the (intra-firm) LV innovation potential.

As stated above, for all types of firms analyzed in Chapter 6, the LV *REGPUSH* was *negatively* related to the intra-firm LV *innovation potential*, while the LV *REGPULL* was *positively* related to this intra-firm LV. Given these considerations, the *'positive'* group will - in general - contain those COROP regions which have a relatively *low* (i.e., negative) *score* on the LV *REGPUSH* and/or a *high* (positive) *score* on the LV *REGPULL* (this will be further illustrated below). The opposite inferences can be made with respect to the 'negative' group of regions.

From the analysis of the previous chapter, it appeared that the impact of both regional LVs on the intra-firm LV innovation potential *differ* in (absolute) value.[6] Consequently, in case the impact of one specific regional LV is clearly dominant - i.e., the REGPUSH impact for the small 'old line' and 'technologically promising' firms - we will especially *concentrate* on the regional scores with respect to this *'dominant'* LV. Concerning the small 'new line' firms attention will focus on *both* regional LVs however, as in this case the impact of the LV REGPUSH appeared to be less dominant than for the type a and c firms discussed in the previous chapter (see the PLS results presented in Tables 6.2, 6.4 and 6.6).

Given the above remarks, the following criteria will be used for selecting 'positive' and 'negative' COROP regions:

1. In order to further illustrate the empirical results found in the previous chapter, *first of all* COROP regions will be selected for which the case values with respect to the LV *REGPUSH* differed considerably from *zero*.[7] As stated above, with respect to the small 'new line' firms also 'extreme' regional scores on the LV REGPULL will be considered in selecting 'positive' and 'negative' regions.[8] Given the (sign of the) impact of both regional LVs on the intra-firm LV innovation potential, COROP regions having (strong) *negative* case values for the *LV REGPUSH* and/or (strong) *positive* case values[9] for the *LV REGPULL* will be placed in the 'positive' group of regions. On the other hand, the 'negative' group will consist of those COROP regions having (strong) *positive* case values for the LV *REGPUSH* and/or (strong) *negative*[10] case values for the LV *REGPULL*.

2. Secondly, COROP regions not selected on the basis of their 'extreme' case values with respect to the LV REGPUSH (and/or REGPUSH for the small 'new line' firms), but *(also) performing relatively (un)favourable* as regards the performance of its firms on the intra-firm LV innovation potential will be selected. In order not to select such COROP regions on the basis of a very small number of observations, as regards this second criterion only COROP regions in which the *number of observations is larger than five* will be considered. As will be clear, COROP regions selected on the basis of this second criterion will be added to the 'positive' or 'negative' group, in case of a relatively favourable or unfavourable performance of its firms concerning the LV innovation potential, respectively.

3. Thirdly, we will also present the empirical results for the three large metropolitan regions mentioned before - viz., Amsterdam, Rijnmond and The Hague. So, even when these regions are not selected on the basis of the two criteria mentioned above, the 'performance' of (the firms in the sample located in) these regions will be analyzed. In this case they will be placed in the 'positive' or 'negative' group when the average case values of its firms concerning the LV innovation potential is larger or lower than zero, respectively.

In the next sections COROP regions will be selected on the basis of the three general criteria mentioned above. For both the 'positive' and 'negative' regions selected also the case values - of the firms (in the inquiry) located in these regions - with respect to the firm-specific LV *innovation potential* will be presented. More specific results - also concerning the exact specification of the criteria mentioned above - will be presented in the next sections. We will start with the small 'old line' firms.

7.3 SPATIAL VARIATIONS IN THE INNOVATION POTENTIAL OF SMALL 'OLD LINE' FIRMS

In this section we will concentrate on the *innovation potential* of the *small 'old line' firms* in the sample discussed in the previous chapter. It can be derived from the final 'restricted' PLS model presented in Table 6.2 that - as far as *regional* dimension is concerned-especially the *negative* impact of the LV *REGPUSH* appeared to be important. Consequently, as regards the first criterion discussed in the previous section, we will select those COROP regions which appeared to have rather 'extreme' case values with respect to this regional LV.[11] Given the fact that the estimated inner relation between the LVs REGPUSH and innovation potential is *negative* (see Table 6.2), a COROP region will be placed in the '*positive*' group when:

* The case value for the LV REGPUSH - concerning its type 1 areas (see the foregoing section) - is smaller than or equal to -1.

In a similar vein, a COROP region will be added to the '*negative*' group when:

* The case value for the LV REGPUSH - concerning its type 1 areas - is larger than or equal to 1

Next, as regards the second criterion discussed in the previous section, COROP regions *not* selected on the basis of their case values for the LV *REGPUSH*, but (also) having a rather (un)favourable performance as regards the endowment its firms (in the sample) with innovative capacity will be selected. In this context, COROP regions- with a number of observations larger than five - for which the average score of its firms concerning the intra-firm LV innovation potential is larger than or equal to 0.3 - or lower than or equal to -0.3 - will be added to the 'positive' or 'negative' group, respectively.[12]

As stated before, we will also analyze the 'performance' of the three large metropolitan areas separately, even when they are not selected on the basis of the criteria specified above. In this latter case, they will be added to the 'positive' or 'negative' group respectively, if the *average score* of *its firms* with respect to the *intra-firm LV innovation potential* is *positive* or *negative*.

For *each selected COROP region* also the empirical results concerning the *number* of small 'old line' firms in the sample having *positive* or *negative* case values for the (firm-specific) LV *innovation potential* will be presented. In this context, the *aggregate score* of these firms on this LV will be provided for each selected region as well (see also below).

In the tables presented below the selected COROP regions have been classified into three *types of regions* depending on whether these COROP regions belong to the *central, intermediate* or *peripheral zone* of the Netherlands. In this respect - see also Bouwman et al (1985a) and Davelaar and Nijkamp (1987d, 1989e) - the *peripheral zone* will consist of the COROP regions 10 till 90, 120, 140, 310 and 370 till 390.[13] Next, the *intermediate zone* will consist of the COROP regions 100, 110, 130, 150, 160, 180, 190, 280, 320 till 360 and 400. The COROP regions 170, 200 till 270, 290 and 300 on the other hand belong to the *central zone*.

In Table 7.1 the results of the selected 'positive' regions are summarized, while Table 7.2 contains the outcomes for the 'negative' regions.

Table 7.1 Selected 'Positive' Regions for the 'Small Old' Line Firms

	COROP REGION	(1) DUM1	(2) REGPUSH	(3) DUM2	(4) REGPULL	(5) positive inn. pot.	(6) negative inn.pot.	(7) aggregate inn.pot.
P	10. Oost-Gr.	-	-2.4	-	2.1	3	1	8.8
E	40. Nrd.-Fr.	3	-1	-	1.1	5	4	0.2
R	50. Zw.-Fr.	-	-1.5	-	0.7	2	1	2.0
I	60. Zo.-Fr.	-	-1.3	-	1.4	0	3	-2.9
P	80. Zo.-Dr.	-	-2	-	1.5	0	1	-0.2
H	90. Zw.-Dr.	-	-2	-	0.8	3	2	-0.3
E	370. Nrd.Lim.	5	-1	-	1.0	9	4	4.3
R								
	120. Twente	11	-0.7	-	1.2	11	11	8.3
I	160. Zw. Gld.	-	-2.1	-	-0.8	1	3	-2.4
N								
T	330. WN.Brab.	1	-0.2	-	-0.8	10	3	6.0
C								
E	220. Zaanst.	15	-1.1	1	-1.0	7	9	2.1
N	291. Rijnm.	14	-1.0	1	-1.3	10	6	4.0
		impact DUM1 on REGPUSH: +0.5		impact DUM2 on REGPULL +2.2				

Table 7.2 Selected 'Negative' Regions for the Small 'Old Line' Firms

COROP REGION		(1) DUM1	(2) REGPUSH	(3) DUM2	(4) REGPULL	(5) positive inn. pot.	(6) negative inn.pot.	(7) aggregate inn.pot.
P E R	30. Ov.Gron. 70. Nr.Dren.	5 -	1.3 1.8	- -	1.8 1.5	3 0	8 3	-3.0 -2.8
I N T	190. Alkm. 150. Arn.Nij.	3 11	1.3 0.3	3 1	-0.7 -0.1	1 2	5 16	-0.5 -7.7
C E N T R A L	200. IJmond 210. Haarl. 231. Amsterd. 232. Ov.Amst. 250. Leid.Bol. 260. The Hag.	- 6 14 - 1 7	3.3 1.4 1.8 2.5 1.5 -0.2	- - - 2 - 4	-1.0 -1.1 -1.0 -1.0 -1.5 -1.4	0 2 1 0 3 3	2 4 13 7 5 8	-1.9 -1.8 -8.1 -7.0 -2.5 -2.4

Before discussing the empirical results, the following remarks
are in order in interpreting both tables presented above. As can be
derived from Table 6.2, in the (final) PLS model for the small 'old
line' firms, the *DUM1* indicator belongs to the indicator set of the LV
REGPUSH, while the *DUM2* indicator forms part of the indicator set of
the LV *REGPULL*. Consequently, as regards the LV *REGPUSH* this implies
that at the *intra-COROP level* the case values for (firms located in)
'urban' (i.e., DUM1) areas *differ* from the (firms located in) other-
i.e., the type 1 and 2 - areas. These *intra-COROP differences* in the
case values for the LV REGPUSH - applying uniformly to all COROP
regions (see the previous section) - are reported in the *lower part* of
the *second column* of Table 7.1. As can be derived from this column
these case values are 0.5 larger for the 'urban' vis-a-vis the other
areas within *each* COROP region. Also, in the second column of both
tables the *number of cases* in the sample located in the 'urban' areas
of each selected COROP region has been reported.

The *third column* of Tables 7.1 and 7.2 contains the *case values*
of the LV *REGPUSH* for (firms located in) the *type 1 and 2 areas* of each
selected COROP region. Taking the example of COROP region 40 in Table
7.1, the third column reveals that this case value for (firms located
in) the type 1 and 2 areas of this COROP region is equal to -1. Given
the intra-COROP differences discussed above, the case value of this LV

for (firms located in) the 'urban' areas within this COROP region then becomes: -1 + 0.5 = -0.5.[14]

In a similar vein, the case values for the LV *REGPULL* differ for (firms located in) *type 2 areas* vis-a-vis *type 1 and 3 areas* of each COROP region. As can be derived from the *fourth column* of Table 7.1, these case values are *larger* - i.e., with a positive discrepancy of 2.2 - for *type 2 areas* than for the *other areas* within each COROP region. This column also contains the *number of cases* in the sample - within each selected COROP region - located in these *'DUM2' areas*. The *fifth column* of both tables then reports the case values for this LV *REGPULL* - i.e., for type 1 and 3 areas - with respect to each selected COROP region.

Next, the *sixth* and *seventh columns* of both tables contain the *number of small 'old line' firms* in the sample - within each selected COROP region - having *positive* or *negative* case values for the (firm-specific) LV *innovation potential*, respectively. Finallly, *column 8* contains the *aggregate* score of the firms in the sample on this LV for each selected COROP region. As regards the four small 'old line' firms in the sample which are located in COROP region Oost-Groningen (i.e., COROP region 10), *3 firms* have an *above average* - i.e., zero - case value for the intra-firm LV *innovation potential*, while *1 firm* scores *below average* (see columns 6 and 7 of Table 7.1). The aggregate score on this LV - i.e., of these four firms together in this region - is equal to 8.8 (see column 8).

After these interpretative remarks, our main attention will focus upon stressing some *general trends*. First of all, it can be deduced from both tables presented above that - in general - COROP regions having (strong) *negative* case values for the regional LV *REGPUSH* lodge the *more innovation potential* firms. The opposite conclusion holds for those COROP regions having (strong) *positive* case values for this LV. This pattern is of course at the basis of the (estimated) *negative* inner relation between the *LVs REGPUSH* and *innovation potential* found in the previous chapter.

In general, COROP regions having (strong) *negative* case values for the LV *REGPUSH* have *positive* case values for the LV *REGPULL* (see Table 7.1). An *opposite pattern* holds for the COROP regions having (strong) positive scores for the LV REGPUSH (see Table 7.2).

As regards the *types* of COROP *regions* included in the *'positive'* and *'negative'* group, both tables reveal an interesting pattern. In this respect, several of the selected *'positive'* *regions* are located in the *peripheral* zone of the Netherlands. Looking more specifically at the scores on the (intra-firm) LV *innovation potential* it appears that several of these *'peripheral'* COROP regions do indeed perform relatively *favourably*. This conclusion does not hold for *three* 'positive' - viz., the COROP regions 60, 80 and 90 - 'peripheral' regions, however. Although these regions have (strong) *negative* scores with respect to the regional LV *REGPUSH*, the performance of its firms concerning the (firm-specific) LV *innovation potential* is below average. However, even when these three regions are left out of consideration - and in a similar vein COROP region 160 - the *peripheral COROP regions* still make up the *greater part* of the selected *'positive'* regions in Table 7.1.

The West-Noord Brabant and Twente regions (i.e., regions 330 and 120, respectively) have been added to the 'positive' group on the basis of the second criterion. Both these regions also have negative scores on the LV REGPUSH.

On the other hand, a *contrasting spatial pattern* is revealed by Table 7.2, however. *All* COROP regions included in this *'negative'* group perform (very) *poorly* as regards the (firm-specific) LV *innovation potential*. From this table, it can also be concluded that the *greater part* of the selected *'negative'* *regions* consists of COROP regions located in the *central* - i.e., Rimcity - *zone of the Netherlands*. On the other hand however, these *centrally located COROP regions* perform relatively *favourably* with respect to several of the *regional indicators* selected in the previous chapter. This pattern largely 'explains' the empirical results of Chapter 6 in which it was found that rather many of these indicators were *positively* related to the LV *REGPUSH* which by itself was *negatively* related to the (intra-firm) LV *innovation potential*.

As stated above, the 'negative' group largely consists of COROP regions located in the *central zone* of the Netherlands. In this context, 6 out of the 10 'negative' COROP regions in Table 7.2 belong to the central zone, while each of these regions is characterized by a relatively poor performance of its firms on the firm specific LV innovation potential. As a whole, only 9 out of the 48 small 'old line' firms in the sample located in these central 'negative' regions have a positive score on the firm-specific LV innovation potential.

With respect to the three *large metropolitan regions*, only Rijnmond (i.e., sub-COROP region 291) could be placed in the 'positive' set of regions. As regards these metropolitan regions, one of the most interesting results from our analysis is the very *poor performance* of the *Greater Amsterdam Area* (i.e., sub-COROP regions 231 and 232). Only *1 out of 21 firms* of 'old line' in the sample located in this region appears to have a *positive score* on the firm-specific LV *innovation potential*. Consequently, the conclusion seems warranted that the Amsterdam region appears to possess the weakest components - in the sense of their potential to innovate - of the small 'old line' firms.

It is noteworthy in this context, that also the region Arnhem-Nijmegen (i.e., COROP region 150) - added to the 'negative' group on the basis of the second criterion - has a rather poor performance. Only 2 out of 18 small 'old line' firms in the survey located in this region are on the positive side of the firms' case values for the LV innovation potential.

As regards the *intra-regional* (intra-COROP) differences with respect to the LV *REGPUSH*, it has been discussed above that the case values for this LV are *larger* - i.e., less negative or more positive- for (firms located in) the '*urban*' areas than for the 'non-urban' (type 1 and 2) areas within each COROP region. So, given the *negative* impact of the LV *REGPUSH*, it appears that at the intra-regional level firms located in 'urban' areas are in general less well endowed with innovative capacity than firms located in the 'non-urban' areas.[15]

This conclusion even holds when the *three large metropolitan areas* - of which especially Amsterdam and The Hague appear to be poorly endowed with 'innovation potential' firms (see Table 7.2) - are removed from the sample. In this case[16], the (estimated) inner relation between the LVs *REGPUSH* and *innovation potential* equals -0.17, while at the intra-COROP level the 'urban' areas have again the largest - i.e., with a positive discrepancy of 0.45 - case values for the LV *REGPUSH*.[17] Consequently, as regards the small 'old line' firms the more innovation potential areas appear to be *non-central* and (as far as the intra-regional level is concerned) '*non-urban*' in character.

After having discussed the results for the small 'old line' firms, in the next section we will focus on the small 'new line' firms.

7.4 REGIONAL INNOVATION POTENTIAL OF SMALL 'NEW LINE' FIRMS

In this section we will select 'positive' and 'negative' regions with respect to the small 'new line' firms discussed in the previous chapter. As regards these firms, it can be concluded from Table 6.4 that the *negative* impact of the LV *REGPUSH* - upon the (intra-firm) LV *innovation potential* - is again more important than the *positive* impact of the LV *REGPULL*, though *less dominant* than for the small 'old line' firms. In the final ('restricted') PLS model for these firms the estimated inner relation between the LV REGPUSH and the LV innovation potential equals -0.09, while the impact of the LV REGPULL on the latter (intra-firm) LV is equal to 0.06.

Consequently, as regards the *first* selection criterion (see section 7.2) *both* the case values for the regional LVs REGPUSH and REGPULL will be considered in selecting 'positive' and 'negative' regions. In this respect, it can be derived from Table 6.4 that for these firms *both* the *DUM1* and *DUM2* indicator belong to the set of regional indicators of the LV *REGPUSH*. Consequently, concerning this regional LV the case values will differ for the type 1, 2 and 3 areas within each COROP region. On the other hand, the case values for the LV REGPULL is the same for - all three types of areas within - each COROP region.

Given the above remarks, as '*positive*' regions will be selected those COROP regions for which the *combined* impact of the scores on both regional LVs - i.e., for type 1 areas within a COROP region (as far as the case value for the LV REGPUSH is concerned) - is *larger than 0.1*. So, in the '*positive*' group will be included those COROP regions for which the *following condition holds*:

$$-0.09 * \text{CV REGPUSH}_{R1} + 0.06 * \text{CV REGPULL}_R > 0.1 \qquad (1)$$

in which,

CV REGPUSH_{R1} = estimated case value for the LV REGPUSH with respect to the type 1 areas within COROP region R

CV REGPULL_R = estimated case value for the LV REGPULL with respect to COROP region R. Given the fact that both the DUM1 and DUM2 indicator do *not* belong to the regional set of indicators of this LV, this case value is the same for (all areas within) COROP region R.

In a similar vein, in the *'negative'* group of regions will be included those COROP regions for which the *following condition holds*:

$$-0.09 * \text{CV REGPUSH}_{R1} + 0.06 * \text{CV REGPULL}_R < -0.1 \qquad (2)$$

As regards the *second criterion* discussed in the previous sections, the COROP regions selected on the basis of their ('extreme') case values for the regional LVs will again be supplemented with those regions which (also) have rather 'extreme' scores of its firms concerning the firm specific LV innovation potential. In this context, the *same criterion* as for the small *'old line' firms* will be used in (further) selecting 'positive' and 'negative' regions (see section 7.3). Also with respect to the small 'new line' firms the results for the three large metropolitan areas will be presented, even when they are not selected on the basis of the criteria discussed above.

Table 7.3 below contains the results for the selected 'positive' regions, while the outcomes for the 'negative' regions are summarized in Table 7.4. For the interpretation of both tables, the reader is referred to the previous section.

Table 7.3 Selected 'Positive' Regions for the Small 'New Line' Firms

	COROP REGION	(1)\nDUM1	(2)\nREGPUSH	(3)\nDUM2	(4)\nREGPULL	(5)\npositive\ninn. pot.	(6)\nnegative\ninn.pot.	(7)\naggregate\ninn.pot.
P\nE\nR\nI	370. Nrd.Limb.	6	-1.4	-	0.0	18	13	8.0
	380. Mid.Limb.	-	-1.4	-	0.2	11	7	3.7
	60. Zo.-Fr.	-	0.0	-	0.1	4	2	3.5
	80. Zo.-Dr.	-	0.1	-	0.5	7	4	4.2
I\nN\nT\nE\nR\nM	110. Zw. Over.	6	-0.8	-	1.7	4	5	1.7
	330. WN. Brab.	6	-0.8	1	2.5	17	17	10.4
	351. Stadgw DB	7	-0.8	1	1.7	8	3	5.4
	352. Ov.NO.Nbr	-	-1.0	-	1.7	8	8	4.1
	400. ZIJP	-	-1.3	-	0.0	2	2	1.2
	320. Ov. Zeel.	-	1.3	-	2.0	5	3	4.4
C\nE\nN	260. The Hag.	7	0.2	1	-0.9	3	7	2.9
	270. Delft+W.	5	-0.5		-0.9	9	5	6.3
		impact\nDUM1 on\nREGPUSH:\n+0.25		impact\nDUM2 on\nREGPUSH\n-0.2				

Table 7.4 Selected 'Negative' Regions for the Small 'New Line' Firms

	COROP REGION	(1)\nDUM1	(2)\nREGPUSH	(3)\nDUM2	(4)\nREGPULL	(5)\npositive\ninn. pot.	(6)\nnegative\ninn.pot.	(7)\naggregate\ninn.pot.
P\nE\nR\nI	20. Delfzijl	-	1.6	-	-0.6	2	2	0.7
	30. Ov. Gron.	12	-0.6	-	0.4	5	13	-6.7
I\nN\nT\nE\nR	280. Oost. ZH.	-	0.7	6	-1.0	6	14	-5.2
	180. Kop NH.	-	-0.2	-	-1.6	4	7	-3.9
	340. Midden NB	11	-0.9	1	0.2	8	21	-9.8
C\nE\nN\nT\nR\nA\nL	200. Ijmond	4	0.9	-	-1.6	3	4	-1.6
	210. Agg.Haarl	7	0.7	3	-1.3	4	6	-0.7
	231. Amsterdam	23	2.4	-	-0.8	6	17	-6.1
	232. Ov. Gr.Am	-	3.3	6	-0.8	6	10	-5.5
	240. Gooi+V.	1	0.7	4	-1.9	3	4	-1.3
	291. Rijnmond	32	1.0	11	-0.6	16	29	-9.9
	292. Ov. Gr.R	-	1.4	8	-0.6	4	5	-2.3
	300. Zo. ZH.	3	-0.7	2	0.0	5	20	-7.6

It can be concluded from the second and third column of Table 7.3 that at the intra-regional (i.e., intra-COROP) level the discrepancies in the case values for LV *REGPUSH* i.e., concerning the type 1, 2 and 3 areas within each COROP region - are rather small. Within each COROP region the case values for the 'urban' (i.e., DUM1) areas are +0.25 larger - and for the type 2 (i.e., DUM2) areas -0.2 smaller - than for the type 1 areas (see also below).

It can be concluded from both tables that *all 'positive' regions*-selected on the basis of their *negative* case value for the LV *REGPUSH* and/or their *positive* case value for the LV *REGPULL* - (indeed) perform relatively *favourably* as regards the (firm-specific) LV *innovation potential*. On the other hand, all 'negative' regions - with the exception of Delfzijl (i.e., COROP region 20) - selected on the basis of their *positive* case value with respect to the LV *REGPUSH* and/or their *negative* case value with respect to the LV *REGPULL* have rather *unfavourable scores* of its firms concerning this intra-firm LV.

Concerning the 'positive' group, half of the selected regions belongs to the *intermediate zone* of the Netherlands. In this context, one of the most interesting results of this analysis is the rather favourable performance of the firms located in the Southern parts of the Netherlands. The COROP regions 330, 351, 352, 370 and 380 are all located in the Southern provinces of Noord-Brabant and Limburg.

On the other hand however, COROP regions located in the *central zone* of the Netherlands are again *weakly represented* in the 'positive group*. In this context only *2* COROP regions included in this group belong to this central zone. Although not selected on the basis of the first two criteria specified above, the Metropolitan Region The Hague (i.e., COROP 260) has been placed in this group because of the *average positive* score of its firms (in the sample) on the LV *innovation potential*. As can be concluded from columns 6 and 7 of Table 7.3 however, this favourable average score is caused by a few (i.e., 3) firms having a relatively high innovation potential. The greater part of the small 'new line' firms in the sample - i.e., 7 out of 10- located in this region has a *negative score* on this intra-firm LV.

The centrally located region Delft and Westland (i.e., region 270) has been included in Table 7.3 on the basis of the second criterion specified above. This also holds for the 'peripheral' regions Zuidoost Friesland and Zuidoost Drente (i.e., regions 60 and 80) and the 'intermediate' region Overig Zeeland (i.e., region 320).

With respect to Table 7.4, the *centrally located COROP* regions again make up the *greater part* of the selected *'negative' regions*. In this context, *8 out of the 13 regions* included in this table are located in the central zone of the Netherlands. As stated before, these centrally located regions have rather favourable scores with respect to several of the regional indicators selected in the previous chapter.

This pattern of course largely explains the empirical findings of the previous chapter. In that chapter it was found that rather many of the selected regional indicators were *positively* related to the LV *REGPUSH* which by itself was *negatively* linked to the LV *innovation potential*. From column 4 of Table 7.4 it can be concluded that 7 out of the 8 central COROP regions included in this table have (strong) *positive* case values for the LV *REGPUSH*. An opposite pattern holds for their case values with respect to the LV *REGPULL* - which is positively related to the intra-firm LV innovation potential - however.

Both *Amsterdam and Rijnmond* (i.e., sub-COROP regions 231 and 291) - and their immediate surroundings (i.e., sub-COROP regions 232 and 292) - clearly belong to this *negative set of regions*. So, although Rijnmond performed relatively favourably with respect to the small 'old line' firms, this region appears to be less well endowed with small 'new line' firms having a relatively high (intra-firm) potential to innovate. The Amsterdam region on the other hand, belongs to the 'negative' group *both* for the small 'old line' and 'new line' firms. In a *relative* sense however, the performance of this region concerning the small 'new line' firms appears to be *less poor* than for the small 'old line' firms (see Tables 7.2 and 7.4).

The regions Overig Groningen, Kop van Noord Holland, Zuidoost Zuid-Holland and Midden Noord Brabant (i.e., regions 30, 180, 300 and 140) have been added to the 'negative' group on the basis of the second criterion mentioned above.

As stated above, the intra-regional differences concerning the scores on the LV REGPUSH are rather small. The 'urban' areas within each COROP region have (slightly) larger - i.e., less negative or more positive - case values for the LV *REGPUSH* than the other (type 1 and 2) areas.

However, when the three large metropolitan regions are removed from the sample - see Table 6.5 - the *DUM1* indicator becomes *positively* related to the LV *REGPULL*.[18] In this case, also the *positive* impact of this latter LV upon the firm specific LV *innovation potential* doubles- i.e., from a value of 0.05 to 0.10 - while the 'urban' areas within each COROP region have considerable *larger scores* - i.e., with a positive discrepancy of 0.7 - on the LV *REGPULL* than the other 'non-urban' areas. This indicates that - apart from the three large metropolitan areas - at the *intra-regional* (i.e., intra-COROP) level firms located in the 'urban' areas are - in general - *better* endowed with *innovative capacity* than firms located in the 'non-urban' areas.

Consequently, as a general conclusion we find that especially the Southern parts of the Netherlands appear to be well equipped in terms of the innovation potential of small 'new line' firms. On the other hand, several centrally located COROP regions - including the Amsterdam and Rijnmond region - again appear to perform relatively unfavourably.

In the next section the results for the small 'technologically promising' firms will be presented.

7.5 REGIONAL INNOVATION POTENTIAL OF SMALL 'TECHNOLOGICALLY PROMISING' FIRMS

In this section the small firms in the sample expected to have rather favourable technological prospects - i.e., on the basis of the selected SIC-codes by the Netherlands Economic Institute (see the previous chapter) - will be briefly considered. In this case the number of observations is relatively small - i.e., compared to the number of cases for the small 'old line' and 'new line' firms. Consequently, one should be (more) careful in interpreting the results with respect to individual COROP regions.

As can be derived from Table 6.6, for these firms the LV *REGPUSH* is again (strong) *negatively* related to the LV *innovation potential*, while the impact of the LV *REGPULL* upon this intra-firm LV equals zero.

Consequently, as regards the *first* criterion for selecting 'positive' and 'negative' regions we will consider regions having 'extreme' case values for the LV *REGPUSH*. In this context, the *same* (first) *criterion* as for the small *'old line'* firms will be applied in selecting 'positive' and 'negative' COROP regions. This also holds for the *second* (and third) criterion for further selecting 'positive' and 'negative' regions (for details, see section 7.3).

On the basis of these criteria the following 'positive' and 'negative' regions have been included in Tables 7.5 and 7.6 below, respectively.

Table 7.5 Selected 'Positive' Regions for the Small 'Technologically Promising' Firms

COROP REGION	(1) DUM1	(2) REGPUSH	(3) DUM2	(4) REGPULL	(5) positive inn. pot.	(6) negative inn.pot.	(7) aggregate inn.pot.
P E R 370. Nrd.Limb.	3	-1.9	-	0.2	5	2	3.6
380. Mid.Limb.	-	-1.9	-	0.4	3	0	4.1
I N T E R 330. WN. Brab.	-	-1.0	-	0.6	3	3	0.3
340. Mid. NB.	1	-1.1	-	0.1	3	1	1.7
351. Stadg.DB	1	-1.2	1	0.0	2	1	0.9
352. Ov.NO.NB	-	-1.2	-	0.0	3	1	2.4
150. Arnh.Nij.	5	-0.4	-	0.1	4	2	2.1
	impact DUM1 on REGPULL: +1.4		impact DUM2 on REGPUSH +0.4				

Table 7.6 Selected 'Negative' Regions for the Small 'Technologically Promising' Firms

COROP REGION	DUM1	REGPUSH	DUM2	REGPULL	positive inn. pot.	negative inn.pot.	aggregate inn.pot.
P E R 30. Ov. Gron.	2	0.1	-	0.6	2	5	-3.3
I N T 130. Veluwe	1	0.2	-	-0.1	2	8	-4.9
C E N T 231. Amsterdam	11	2.4	-	-1.3	3	8	-5.3
232. Ov. Gr.Am	-	2.9	5	-1.3	2	6	-4.1
260. The Hag.	2	0.6	1	-1.6	1	3	-2.5
291. Rijnmond	15	-0.2	5	-1.6	9	14	-2.0

Looking more specifically at the intra-regional (i.e., intra-COROP) differences it appears - see Table 6.6 - that the *DUM2* indicator belongs to the regional indicators of the LV *REGPUSH* while the *DUM1* indicator belongs to the indicator set of the LV *REGPULL* (which by itself is unrelated to the LV innovation potential).

Consequently, at the intra-regional level the case values for the LV REGPUSH are the same for the type 1 and 3 areas (see section 7.2), while the 'DUM2' areas have larger - i.e., with a positive discrepancy of 0.4 (see column 2 of Table 7.5) - scores for this LV.[19]

Also in this case COROP regions having (strong) *negative* scores for the LV REGPUSH do indeed perform relatively *favourably* as regards the intra-firm LV *innovation potential*. The opposite conclusion holds for those regions having (strong) *positive* case values for the LV REGPUSH.

As can be derived from Table 7.5, the greater part of the selected 'positive' regions belongs to the intermediate zone of the Netherlands. Also in line with the results of the foregoing section appears to be the rather *favourable* performance of the Southern parts of the Netherlands. In this context, 6 out of the 7 'positive' COROP regions selected are located in the Southern provinces of Noord-Brabant and Limburg. Although the Arnhem-Nijmegen region (i.e., region 150) appeared to perform rather poorly with respect to the small 'old line' firms, as regards the 'technologically promising' firms this region could be added - on the basis of the second criterion - to the 'positive' group.

Also concerning the firms analyzed in this section several centrally located (metropolitan) regions again belong to the 'lagging' set of regions. It is noteworthy however, that the number of 'central' regions included in Table 7.6 is clearly smaller than for the other types of firms analyzed in this chapter and only consists of the large metropolitan areas mentioned before. In this context, both Amsterdam (i.e., sub-COROP region 231) and its immediate surroundings (i.e., sub-COROP region 232) again belong to the 'negative' set of regions.

Given the (average) unfavourable performance of its firms with respect to the LV innovation potential both Rijnmond and The Hague (i.e., regions 291 and 260) have also been included in this group. This also holds for the regions Oost-Groningen and Veluwe (i.e., regions 30 and 130, respectively).

Contrary to the *intra-regional results* found in the previous sections - i.e., when the three large metropolitan regions are included - as regards the small 'technologically promising' firms the 'urban' areas do *not* have the largest scores on the LV REGPUSH (which is *negatively* related to the firm specific LV innovation potential).

After *removing* the (firms in the sample located in the) three large metropolitan regions however - see Table 6.7 - the *impact* of the LV *REGPULL* upon the intra-firm LV innovation potential *increases* considerably (i.e., from 0.0 to 0.14).[20] In this case, there are also considerable intra-regional discrepancies with respect to the LV *REGPULL*. In this context, the 'urban' areas within each COROP region have significantly *larger* - i.e., with a *positive* discrepancy of 2.1- scores on this LV than the other 'non-urban' (type 1 and 2) areas.

So just like we found with respect to the 'new line' firms, after removing the three large metropolitan areas the positive impact of the LV REGPULL upon the LV innovation potential increases while at the intra-regional level 'urban areas' have considerable larger scores on the former LV. Consequently, it appears that - apart from the three large metropolitan areas - at the *intra-regional* level firms located in 'urban' areas are in general *better endowed* with innovative capacity than firms located in the other 'non-urban' areas. It is noteworthy however, that these patterns are even more marked for the small 'technologically promising' than for the small 'new line' firms.

In concluding this section we can state that concerning the small 'technologically promising' firms the Southern part of the Netherlands appears to be favourably endowed as regards firms having a relatively high potential to innovate. On the other hand, the three large metropolitan areas (again) appear to belong to the lagging category. These patterns are largely in accordance with the general trends identified in the previous section concerning the small 'new line' firms.

7.6 CONCLUSION

As discussed in previous chapters (see Chapters 1 and 5), the employment generating capacity of the manufacturing sector has declined rather drastically in recent decades. Generally speaking, the decline in the (possibilities for the) generation of product innovations had a negative impact on manufacturing employment in these periods. In this context, the generation of cost-efficient (labour-saving) process innovations became increasingly a critical factor.

In this respect, our third hypothesis - related to the theoretical framework developed in the first part of this study- states that in such later phases of the life cycles of industries and technologies non-metropolitan or non-central regions will become dominant concerning the (further) innovative activities performed.

Consequently, in order to test this hypothesis, in the present chapter we have analyzed (more general) spatial innovation patterns for the Dutch manufacturing sector. In this respect, we have analyzed the performance of Dutch (types of) regions - i.e., on the basis of the industrial survey discussed in Chapter 6 - as regards their *indigenous innovation potential* for the *three types* of small manufacturing firms distinguished in the previous chapter.

To these purposes - and also with the aim of further illustrating the empirical results found in the previous chapter - we first of all selected those COROP regions which had rather 'extreme' scores (i.e., case values) on the regional LVs *REGPUSH* and *REGPULL* as discussed in Chapter 6. By means of this criterion it has been shown that - in general - firms located in such regions do indeed have *relatively favourable* or *unfavourable scores* on the firm specific LV *innovation potential*. Next, we also selected those regions which were not initially selected by their 'extreme' scores on these regional LVs, but (also) appeared to perform relatively (un)favourably as regards the performance of its firms (in the sample) on this firm-specific LV. In this context, the performance of the three large metropolitan regions separately has also been analyzed, even when they were not selected on the basis of the criteria mentioned above.

Several of the spatial innovation patterns detected appear to support our theoretical framework - and in particular the third hypothesis deduced from this framework - developed in the first part of this study.

As regards the *spatial dispersion* of '*positive*' and '*negative*' regions - i.e., regions characterized by a relatively favourable or unfavourable score of their firms on the intra-firm LV innovation potential, respectively - an interesting pattern could be discerned from the data. Concerning *all three types* of firms analyzed in this chapter, several COROP regions located in the *central zone* of the Netherlands appeared to have a relatively *poor performance* of its firms - i.e., their case values - on this LV. In this context, the *greater part* of the selected '*negative*' regions consisted of COROP regions located in the *central zone* of the *Netherlands*. On the other hand however, as regards this firm-specific LV only a *few regions* in the *central zone* could be designated as a '*positive*' region. As stated above, these conclusions apply to *all three types of firms* analyzed in this chapter.

As regards the Amsterdam area, our empirical results suggest that the relative poor performance of this area - i.e., in terms of the development of manufacturing employment in recent decades (see Chapter 5) - is also mirrored by a relative weak innovation potential of its manufacturing firms. In this respect, the *Amsterdam* area belonged to the '*negative*' *set* of regions for *each type of firm* analyzed in this chapter. However, our results also seem to indicate that the relative poor performance of the Amsterdam area concerning the innovation potential of its firms applies especially to the (small) 'old line' firms (these results are largely in accordance with the patterns depicted in Figure 5.5 for this area).

As regards the small 'old line' firms, several COROP regions located in the '*peripheral*' zone of the Netherlands appeared to perform relatively favourably. With respect to these firms the greater part of the selected 'positive' regions belonged to this peripheral zone. This conclusion also holds when those ('peripheral') regions selected on the basis of their 'favourable' scores on the regional LV REGPUSH, but scoring below average as regards the firm-specific LV innovation potential, are left out of consideration.

In this context, 6 out of the 10 regions included in the 'negative' set of regions are located in the central zone of the Netherlands. Especially the Greater Amsterdam Region appeared to possess the weakest components of this type of firms.

Rather interestingly, the *'peripheral orientation'* (of innovation potential small 'old line' firms) also appears to be mirrored at the *intra-regional* level. With respect to the small 'old line' firms the 'urban' (DUM1) areas within a COROP region have the *largest* scores on the LV REGPUSH. Given *negative impact* of this LV upon the firm-specific LV *innovation potential* it appears that also at the intra-regional level firms located in these 'urban' areas are - in general-less well endowed with *'innovation potential'* than firms located in the other 'non-urban' areas. This conclusion also holds when the three large metropolitan areas are removed from the sample.

This 'peripheral' orientation of innovation potential firms applies less to the small *'new line'* firms, however. In this context, half of the selected 'positive' regions belongs to the *intermediate* zone of the Netherlands. In this respect, several COROP regions located in the Southern parts of the Netherlands - i.e., the provinces of Noord-Brabant and Limburg - appeared to be well-endowed with the more innovation potential components of these firms.

Also concerning these firms however, the greater part of the selected 'negative' regions - i.e, 8 out of 13 - again belongs to the central zone of the Netherlands. In this context, both Rijnmond and Amsterdam also belong to this negative set of regions.

At the *intra-regional* level the discrepancies between 'urban' and 'non-urban' areas - as regards their case values for the LV REGPUSH-are smaller than for the small 'old line' industries. After removing the three large metropolitan regions however, the *'urban'* areas within a COROP region have (considerable) *larger scores* on the LV *REGPULL* than the 'non-urban' areas, while also the *positive* impact of the latter LV upon the firm-specific LV innovation potential *increases*. So - apart from the three large metropolitan areas - at the intra-regional level small 'new line' firms located in 'urban' areas in general appear to be (slightly) better equipped with innovative capacity than firms located in the other 'non-urban' areas.

Also concerning the small firms expected to have rather favourable technological prospects, the selected 'positive' regions for the greater part belong to the intermediate zone of the Netherlands. Also concerning these firms, several COROP regions located in the two Southern provinces of Brabant and Limburg were added to the 'positive' group. As regards these firms the number of 'central' regions included in the 'negative' set of regions is smaller than for the other types of firms analyzed in this chapter. All three large metropolitan regions however, were included in this set of regions. Consequently, these spatial patterns are largely in accordance with the spatial patterns found with respect to the small 'new line' firms.

Contrary to the results found for the small 'old line' and 'new line' firms however - i.e., when all regions are included - at the *intra-regional* level the 'urban areas' do *not* have the largest scores on the LV *REGPUSH* (which is negatively related to the firm-specific LV innovation potential). When the three large metropolitan areas are again removed from the sample however, it appears that at the intra-regional level firms located in 'urban' areas are generally better endowed with innovative capacity than firms located in the 'non-urban' areas. These intra-regional differences appear to be more pronounced than the results found for the small 'new line' firms - i.e., after excluding the three metropolitan regions.

In summarizing the results of this and the previous chapter, we can state that with respect to all types of manufacturing firms analyzed in these chapters, several regions located in the central zone of the Netherlands appear to be (clearly) lagging behind as regards their indigenous innovation potential. This conclusion also holds for the three large metropolitan regions as a whole. In this context, our results also suggest that the relative indigenous innovation potential of *peripheral* zones - and/or the 'non-urban' areas within (COROP) regions - increases in particular for those sectors that are linked to former technological 'trajectories' (see also the spatial patterns found for type c, b and a firms respectively). Consequently, the analysis performed in this chapter indicates that the spatial deconcentration of the locus of innovative activities is indeed most pronounced for 'old technologies' as expected on the basis of our theoretical framework and simulation model developed in Chapters 3 and

4. Or, in other words, the consequences of economic and technological restructuring are first noticed in more central or metropolitan areas.

Consequently, the (more general) spatial *innovation patterns* found in this and the previous chapter largely *correspond to the* relatively poor performance of the more *centrally located (metropolitan) regions* as regards employment in the *manufacturing sector* as a whole (see Keeble et al 1983, Wever 1985 and Chapter 5 of this study). Thus, besides the other - non-technological - causal forces identified with respect to the 'urban-rural' shift in manufacturing employment, the *technological dimension* also appears to be important in this context.

As stated above, these results are largely in accordance with our theoretical framework, in particular concerning the third hypothesis on geographical deconcentration of innovative activities in later phases of the life cycles of industries and technologies.

Notes to Chapter 7:

1. As stated in the previous chapter, the inclusion - or analysis -of large firms renders the intra-firms scores on the LVs innovation potential and innovativeness *less comparable*. In the case of type d firms distinguished in the foregoing chapter, for example, a rather *small* proportion of firms has (large) *positive* case values for the intra-firm LV innovation potential, while the *greater* proportion of firms has (small) *negative* case values for this LV. At the regional dimension this implies that a relatively favourable performance is especially determined by this smaller proportion of firms having (large) positive scores on this LV. Given these considerations, in this chapter we will concentrate on the type a, b and c firms discussed in the previous chapter. It should be recalled from the discussion of Tables 6.8 and 6.9 however, that also with respect to type d firms the 'high quality' production environments - i.e., in general the COROP regions located in the Rimcity (see also the results of this chapter)- performed relatively poorly as regards their endowment with (the smaller proportion of) firms having favourable scores on the LV innovation potential.

2. For details, see the previous chapter.

3. The scores - i.e., case values - for the (regional and intra-firm) LVs which will be analyzed (and presented) in this chapter are based on the final 'restricted' PLS models discussed in Chapter 6.

4. As in this case both the DUM1 and DUM2 indicator do *not* belong to the set of regional indicators used to approximate the LV *REGPULL*, the (estimated) case value concerning this LV applies uniformly to (all areas within) a specific COROP region.

5. As regards the interpretation of the empirical results to be presented in the next sections, it should be remembered that the case values of all - intra-firm and regional - LVs have unit variance and mean zero.

6. See the estimated inner relations a and y - cf., Figure 6.3 - in the final 'restricted' PLS models presented in Chapter 6.

7. As regards the interpretation of the empirical results presented in the next sections, it should be remembered that the case values of all- intra-firm and regional - LVs (across all firms and regions) have unit variance and mean zero.

8. When the DUM1 and/or DUM2 indicator belong to the set of regional indicators of these regional LVs, selection will be made on the basis of 'extreme' scores with respect to the type 1 areas within each COROP region (see also the next sections).

9. In the case of the small 'new line' firms.

10. See footnote 6 of the present chapter.

11. However, as regards the selected COROP regions also the estimated case values with respect to the LV REGPULL will be presented in Tables 7.1 and 7.2 below. Appendix 1 of this chapter contains the (estimated) case values for both the LV REGPULL and REGPUSH for all COROP regions and all types of firms discussed in this chapter.

12. The average score (i.e., in Tables 7.1, 7.3 and 7.5 jointly) of the *'positive'* regions - selected on the basis of their 'extreme' case values for the *regional LV's* - with respect to this *firm-specific LV* is (also) approximately equal to 0.3. In a similar vein, the average score for the *'negative'* regions (i.e., in Tables 7.2, 7.4 and 7.6 jointly) - selected on the basis of their cases values for both *regional LVs* - is approximately equal to -0.3.

13. These are the COROP-codes of the various COROP-regions: see also the appendices of this and the previous chapter.

14. From columns 2, 4 and 6 and 7 of Table 7.1 it can be derived that 3 'old line' firms in the sample are located in 'urban' areas within this COROP region, while 6 firms are located in type 1 areas of this COROP.

15. Concerning these type 1 and 2 areas, especially the 'DUM2' (type 2) areas - given their larger case values with respect to the LV REGPULL (see column 4 of Table 7.1) - in general tend to perform rather well.

16. See also Table 6.3 of Chapter 6.

17. In this case, the DUM1 areas within each COROP region have again larger - i.e., with a positive discrepancy of 2.4 - case values for the LV REGPULL.

18. In this case, the DUM2 areas within each COROP region have again (slightly) lower - i.e., with a negative discrepancy of -0.3 - case values for the LV REGPUSH than the other (type 1 and 2) areas.

19. Given the negative impact of this LV upon the firm specific LV innovation potential, it appears that at the intra-regional level firms located in the DUM2 areas are - in general - less well endowed with innovative capacity than firms located in the other types of areas.

20. In this case the DUM2 indicator appeared to be unrelated to both the LV REGPULL and REGPUSH.

Appendix 1   Regional Case Values for the LVs REGPUSH and REGPULL
(including the three metropolitan areas)

COROP REGION	'old line' firms		'new line' firms		techn. prom. firms	
	REGPUSH[1]	REGPULL[2]	REGPUSH[3]	REGPULL[4]	REGPUSH[5]	REGPULL[6]
10. Oost Groningen	-2.4	2.1	0.3	-0.8	0.5	0.5
20. Delfzijl e.o.	-	-	1.6	-0.6	0.4	0.6
30. Overig Groningen	1.3	1.8	-0.6	0.4	0.1	0.6
40. Noord Friesland	-1.0	1.1	0.0	0.6	0.2	0.4
50. Zuidw. Friesland	-1.5	0.7	-0.1	0.5	0.1	0.6
60. Zuido. Friesland	-1.3	1.4	0.0	0.1	0.2	0.5
70. Noord Drenthe	1.8	1.5	-0.5	0.3	-	-
80. Zuido. Drenthe	-2.0	1.5	0.1	0.5	0.2	0.6
90. Zuidw. Drenthe	-2.0	0.8	-0.1	0.6	-0.2	0.4
100. Noord-Overijssel	-0.7	0.1	0.4	1.0	-0.5	0.3
110. Zuidw.Overijssel	-0.3	0.1	-0.8	1.7	-0.9	0.0
120. Twente	-0.7	1.2	0.6	-0.5	-0.1	0.0
130. Veluwe	-0.2	-0.1	0.6	0.0	0.2	-0.1
140. Achterhoek	-0.8	0.2	-0.6	-0.3	-0.8	-0.1
150. Arnhem Nijmegen	0.3	-0.1	0.1	0.5	-0.4	0.1
160. Zuidw. Gelderland	-2.1	-0.8	-0.4	0.8	-0.6	0.2
170. Utrecht	0.7	-1.2	0.2	-1.3	0.3	-1.6
180. Kop v. Nrd. Holland	0.8	0.3	-0.2	-1.6	0.2	-1.0
190. Alkmaar e.o.	1.3	-0.7	-0.7	-1.4	0.0	-1.3
200. IJmond	3.3	-0.9	0.9	-1.6	-0.3	-1.6
210. Aggl. Haarlem	1.4	-1.1	0.7	-1.3	0.4	0.0
220. Zaanstreek	-1.1	-1.0	-0.1	-1.0	0.0	0.0
231. Amsterdam	1.8	-1.0	2.4	-0.8	2.4	-1.3
232. Over. Gr. Amsterdam	2.5	-1.0	3.0	-0.8	2.9	-1.3

COROP REGION	'old line' firms		'new line' firms		techn. prom. firms	
	REGPUSH[1]	REGPULL[2]	REGPUSH[3]	REGPULL[4]	REGPUSH[5]	REGPULL[6]
240. Gooi en Vechtstr.	0.7	-0.8	0.7	-1.9	0.2	-1.7
250. Leiden en Bollenstr	1.5	-1.5	-0.7	-0.3	0.2	-1.6
260. Den Haag	-0.2	-1.4	0.2	-0.9	0.6	-1.6
270. Delft en Westl.	0.6	-1.2	-0.5	-0.9	0.3	-1.6
280. Oost Zuid Holl.	0.0	0.6	0.7	-1.0	0.1	-1.6
291. Rijnmond	-1.0	-1.3	1.0	-0.6	-0.2	-1.6
292. Ov. Rijnmond	-0.7	-1.3	1.4	-0.6	0.4	-1.6
300. Zuido. Zuid-Holl.	-0.5	-1.1	-0.8	0.0	-0.7	-1.4
310. Zeeuws-Vlaanderen	0.4	1.3	1.4	1.4	0.0	0.6
320. Overig Zeeland	0.9	0.5	1.3	2.0	-0.1	0.6
330. West Nrd. Brabant	-0.2	-0.8	-0.8	2.5	-1.0	0.6
340. Midden Nrd. Brabant	-0.9	-0.3	-0.9	0.2	-1.0	0.1
351. Stadg. Den Bosch	-0.5	-0.6	-0.8	1.7	-1.2	0.0
352. Ov. Nrdo. Nrd. Brab	-0.4	-0.6	-1.0	1.7	-1.2	0.0
360. Zuido. Nrd. Brabant	0.0	0.0	0.0	0.3	-0.6	0.0
370. Noord-Limburg	-1.0	1.0	-1.4	0.0	-1.9	0.2
380. Midden Limburg	-0.6	0.9	-1.4	0.2	-1.9	0.4
390. Zuid-Limburg	0.3	1.6	-0.7	0.0	-0.2	0.2
400. ZIJP	0.2	-0.1	-1.3	0.0	-0.7	0.0
	impact DUM1 on REGPUSH: +0.5	impact DUM2 on REGPULL: +2.2	impact DUM1 and DUM2 on REGPUSH: +0.25 and -0.2 respectively		impact DUM2 on REGPUSH: +0.4	impact DUM1 on REGPULL: +1.4

Legend:

1. For (firms located in) type 1 and 2 areas of a COROP region.
2.   ,,   type 1 and 3 areas of a COROP region.
3.   ,,   type 1 areas of a COROP region.
4.   ,,   all three types of areas within a COROP region.
5.   ,,   type 1 and 3 areas of a COROP region.
6.   ,,   type 1 and 2 areas of a COROP region.

- = no observation.

8 Generation of product and process innovations

8.1 INTRODUCTION

In this chapter the generation of product and process innovations in Dutch manufacturing establishments will be considered. For this purpose we will use the results of the industrial survey data - among 1,842 main establishments (labelled firms here) of the Dutch manufacturing sector - as discussed in Chapter 6.

Concerning the generation of product and process innovations the impact of various intra-firm variables - like firm size, sector, external and internal R&D efforts - will be analyzed. Besides these intra-firm variables however, also the *spatial* dimension will be taken into account. This spatial dimension is important for the empirical testing of our dynamic incubation framework - and in particular the fourth hypothesis deduced from this incubation framework - as developed in Chapter 3. According to this framework the locus of (further) innovative activities will shift from central to non-central areas in later phases of the life cycles of industries and technologies. As regards the *types of innovations* generated in these later phases however, these central areas are expected to lag behind especially concerning the generation of *process innovations*, as these innovations become increasingly important in later phases of technological progress (see also Porter 1980 and Chapters 1 and 3).

Thus in our framework we (also have to) call attention to the role of innovative activities in explaining the urban-rural shift of manufacturing employment as observed in all EEC countries (see Keeble et al 1983). The empirical results presented in the previous chapters suggest that the rather poor performance of several (metropolitan) regions in the central zone of the Netherlands - as regards the development of manufacturing employment - (indeed) appears to be also mirrored by a rather weak *innovation potential* concerning several types of manufacturing firms. In the present chapter however, we will particularly concentrate on the regional performance concerning the *types* of (i.e., product versus process) of *innovations generated*.

The impact of the intra-firm and spatial variables mentioned above will be analyzed by using a multivariate logit analysis. In this analysis multiple (categorical) variables related to our survey data are simultaneously taken into consideration as explanatory variables for the generation of product and process innovations in Dutch manufacturing establishments. In section 8.2 our analytical framework will be described more in detail.

Concerning the *spatial dimension*, we will distinguish various types of regions in the Dutch context. First, in line with the analysis discussed in the previous chapter, in section 8.3 we will consider the performance of the central, intermediate and peripheral zone as regards the generation of both product and process innovations.

Besides this *regional level* of analysis however, the performance of *urban* vis-a-vis *non-urban* areas will be analyzed in section 8.4. Consequently, in this section the generation of product and process innovations in various (types of) Dutch municipalities will be analyzed. Section 8.5 contains some retrospective and prospective concluding remarks.

8.2 ANALYTICAL FRAMEWORK

As indicated in the introduction, our analytical framework will be based on data concerning an extensive (postal) survey among nearly 1,850 main establishments in the Dutch manufacturing sector (see also Chapter 6). These data refer to the innovative activities performed within these establishments in 1983. Details on this survey, held in 1984, can be found in Kleinknecht (1987b). Our estimation results, to be presented in the next sections, will be based on the complete sample results.

In the following sections the generation of product and process innovations (which are new[1] for the whole branch of industry in the Netherlands the firm belongs to) within these firms in 1983 will be analyzed. The logit models (see below) will be estimated separately for product and process innovations. For this purpose, we have selected the following explanatory (categorical) variables for innovative behaviour:

1. The *sector* to which the firm belongs. In line with the previous
   mentioned innovation life cycle concepts (implying a shift from
   product to process innovations), we have subdivided the firms in
   our sample into '*old line*' and '*new line*' firms. As stated in
   Chapter 6, the 'old line' firms belong to the two-digit Dutch
   Standard Business Classification (SBI) with codes SBI-20 to SBI-27
   - denoted here as BRAN(1) -, while the 'new line' firms are marked
   by the SBI-codes 28 to 39 - denoted here as BRAN(2).

2. The *size* of the firm. Here a distinction is made between 'large'
   firms with a number of employees larger than or equal to 100 -
   denoted as WERK(2) - and 'small' firms with less than 100 employees
   - denoted as WERK(1).

3. The *use of external R&D activities*. The question here is whether the
   firm has used external R&D (from universities, research institutes,
   other firms) - denoted as ERD(1) - or not - denoted as ERD(2).

4. The *internal R&D activities* of the firm. In this respect, three
   possibilities are distinguished:
   - the firm has its own separate R&D division, denoted here as
     RDA(1);
   - the firm does not have a specific R&D division, but in the year
     under consideration internal R&D activities have been performed
     by other divisions of the firm, denoted here as RDA(2);
   - the firm has not performed any internal R&D efforts in 1983,
     which will be denoted as RDA(3);

5. The *location* of the firm. In this context, our analysis will be performed for various spatial - i.e., regional and urban - subdivisions of the Netherlands. This spatial dimension of innovative behaviour will be discussed more in detail in the next sections.

In order to determine the impact of the above-mentioned (categorical) variables upon the generation of product and process innovations our multivariate logit model[2] - illustrated in a simplified form for only two explanatory variables - has then the following structure:

$$\log \left[\frac{I}{N - I} \right]_{i(k)j(l)} = U + U_i(k) + U_j(l) + U_{ij}(kl) \qquad (1)$$

where:

$\log \left[\dfrac{I}{N - I} \right]_{i(k)j(l)}$ = logarithm of the ratio of the number of innovating firms (I)[3] and non-innovating firms (N - I), if the (categorical) variable i adopts the value k and the (categorical) variable j the value l

U = constant

$U_i(k)$ = main effect with respect to (categorical) variable i and value k

$U_j(l)$ = main effect with respect to (categorical) variable j and value l

$U_{ij}(kl)$ = interaction effect between the categorical variables i and j, if they adopt the values k and l, respectively

As stated before, the logit model specified above will be estimated separately for the generation of product and process innovations. In this context, the GLIM-package (see Baker et al 1978) has been used in estimating the various logit models which will be presented in the next sections.

Our logit model is treated and estimated according to the guidelines presented in the standard book by Bishop et al (1977). In this respect, the relevant parameters concerning the explanatory variables have been estimated under the following restrictions:

$$\sum_{k=1}^{K} U_i(k) = 0; \quad \sum_{l=1}^{L} U_j(l) = 0; \quad \sum_{l=1}^{L} U_{ij}(kl) = 0; \quad \sum_{k=1}^{K} U_{ij}(kl) = 0 \quad (2)$$

The sum total of the parameters over the various categories of all main and interaction effects is equal to zero. Consequently, as regards the sectoral impact, only the estimated parameters concerning the BRAN(1) variable will be presented; with respect to the BRAN(2) variable the opposite values hold. This also applies, mutatis mutandis, to the estimated parameters of the other main and interaction effects which are included in the tables presented in the next sections.

The research strategy for estimating the above mentioned logit model - for both product and process innovations separately - was based on an initial estimation of the model including all main effects, except for the variable 'location'. Besides the estimated parameter values for the main effects with respect to these intra-firm variables 1 to 4 (see above), the estimated parameters divided by their standard errors will also be reported in the next sections. This test variable is asymptotically normal distributed (see Bishop et al 1977). Consequently, it provides information about the statistical significance of these parameters (like t-values in regression analysis).

Next, the extent to which the inclusion of the variable 'location' leads to a better fit of the model will be examined. The estimated parameter values with respect to this variable will also be presented in the next sections.' In order to determine the statistical significance of this latter variable the change in the likelihood ratio G^2 will be compared to the change in the total number of degrees of freedom. The change (i.e., improvement) in this ratio has a chi-square distribution with a number of degrees of freedom equal to the *change* (i.e., loss) in the total number of degrees of freedom resulting from including extra parameters (for details, see Bishop et al 1977). The statistical level of significance of this main effect will also be reported in the tables presented in the next sections.

In addition to these main effects, higher-order interaction effects will only be included when they lead to a significant improvement (at the 10% significance level) of the statistical fit of the estimated model including all main effects. Also with respect to these (significant) interaction effects the estimated parameter values and the statistical level of significance[4] will be recorded in the next sections.

In the next section our estimated logit models for the regional level of analysis will be presented.

8.3 ESTIMATED LOGIT MODEL FOR THE REGIONAL LEVEL OF ANALYSIS

In this section we will consider the generation of both product and process innovations in the central, intermediate and peripheral zone of the Netherlands (see also Davelaar and Nijkamp 1987d and 1989e). These zones have been defined on the basis of standard statistical areas (so-called COROP-areas) which altogether make up the above-mentioned 3 types of regions.[5]

Consequently, in this section the following possibilities will be distinguished with respect to the location of a specific firm (see the fifth explanatory variable of innovative behaviour mentioned in the previous section):

- Location in the central zone of the Netherlands, denoted here as REG(1)
- Location in the so-called intermediate (or halfway) zone, denoted here as REG(2)
- Location in the peripheral zone, denoted as REG(3)

Given this spatial demarcation, the above mentioned logit analysis has been performed. Table 8.1 below presents the results for the generation of product innovations. Table 8.2 provides the empirical results for process innovations.

Table 8.1 Estimated Logit Model with respect to Product Innovations

Variable	Estimate	Statistical Significance[6]
BRAN(1)	-0.24	-3.4
ERD(1)	0.21	3.1
WERK(1)	-0.14	-2.0
RDA(1)	0.97	7.8
RDA(2)	0.71	6.3
REG(1)	-0.03	-
REG(2)	-0.02	-
REG(1) * RDA(1)	0.21	8%
REG(2) * RDA(1)	0.04	-
BRAN(1) * REG(2)	-0.15	8%
BRAN(1) * REG(1)	0.02	-

$G^2 = 48.3$, d.o.f. = 60

Table 8.2 Estimated Logit Model with respect to Process Innovations

Variable	Estimate	Statistical Significance
BRAN(1)	0.10	1,2
ERD(1)	0.33	3.9
WERK(1)	-0.43	-4.7
RDA(1)	0.64	4.7
RDA(2)	0.32	2.7
REG(1)	-0.16	6%
REG(2)	-0.05	-
BRAN(1) * WERK(1)	-0.19	2%
HULP	0.21	3%

$G^2 = 47$, d.o.f. = 62

As stated in the previous section, in the tables which will be presented in this and the next chapter only the 'independent' parameter estimations with respect to all main and interaction effects will be recorded. With respect to the other parameters it should be recalled that the sum total of the parameters over the various categories of all

main and interaction effects - see equation (2) of the previous section-
is equal to zero.

Taking the example of the 'sector' variable, it can be derived
from Table 8.1 above that the estimated parameter for the 'old line'
firms - i.e., BRAN(1) - equals -0.24. Given the above-mentioned
restrictions, the estimated parameter for the 'new line' firms - i.e.,
BRAN(2) - then becomes +0.24.

In discussing the results of our analysis, we will first
concentrate on the impact of the various intra-firm variables, while
the regional dimension will be considered subsequently.

A first conclusion which can be drawn from Table 8.1 above is
that 'new line' firms show a significantly higher innovativeness than
'old line' firms regarding the generation of product innovations that
are new to the whole branch the firm belongs to. Rather interestingly
however, as regards the generation of process innovations an opposite
pattern appears to hold. In this case the estimated parameter with
respect to the 'sector' variable is positive for the 'old line' firms.
However, it appears from the statistically significant interaction
effect BRAN(1) * WERK(1)[7] that especially the *large 'old line'* firms are
oriented towards the generation of process innovations.

Consequently, the effect of 'old line' and 'new line' sectors
shows a mutually contrasting pattern for product and process
innovations. 'New line' firms are more oriented towards the creation of
product innovations, while (large) 'old line' firms are more oriented
toward the generation of process innovations. It goes without saying
that these results clearly support the innovation life cycle concept
discussed in the first part of this book. According to this concept the
generation of (further) product innovations will be especially
important in earlier phases of the life cycles of industries and
technologies, while in later phases the main emphasis of the innovative
activities will increasingly shift towards further improving the
underlying process technologies.

As far as *firm size* is concerned, it is noteworthy that large
firms are clearly more innovative than small firms. This is indicated
by the significant negative estimated parameter values with respect to

variable WERK(1) (i.e., small firms) for both the generation of product and process innovations. However, it should also be noted that the difference in firm size appears to be particularly important for the generation of process innovations. Apparently small firms are better able to hold pace with large firms in terms of the generation of product innovations compared to the generation of process innovations.

When we consider the estimated main and interaction effects in Tables 8.1 and 8.2 above for the following types of firms: (1) small 'old line' firms, (2) small 'new line' firms, (3) large 'old line' firms and (4) large 'new line' firms, we arrive at the following conclusions:

(1) small 'old line' firms appear to perform relatively poorly for both product and process innovations;

(2) small 'new line' firms have a poor performance for process innovations, but a better performance as regards the generation of product innovations;

(3) large 'old line' firms are forerunners in terms of process innovations, but score relatively poorly in terms of product innovations;

(4) large 'new line' firms are forerunners in terms of the generation of product innovations but not for process innovations;

The importance of internal (i.e., RDA(1) and RDA(2)) and external R&D (i.e., ERD(1)) efforts is evident for both the generation of product and process innovations. However, it is interesting to observe here that the difference in the parameter values for RDA(1) and RDA(2) in Tables 8.1 and 8.2 above indicates that internal R&D is particularly important for the generation of product innovations. Consequently, internal R&D appears to be less important for the generation of process innovations than for the generation of product innovations. As can be concluded from both tables above, this conclusion does not apply to the (importance of) external R&D efforts for the generation of both product and process innovations, however.

Before interpreting the empirical results with respect to the *regional* dimension of innovative behaviour, it is noteworthy to call attention to a hitherto neglected but nevertheless important point, viz., the difference in *quality* (in terms of economic effects) of the innovations considered. It is assumed here that in general there exists a positive correlation between the intensity of R&D efforts and the economic consequences (in terms of profitability, market share, relative competitive position) of the innovations generated by firms (or regions). Thus it seems plausible to hypothesize a *positive* link between (internal and external) R&D efforts and the average *quality* of the innovations generated. This would imply that a subdivision of firms which have introduced (product or process) innovations into firms with high and low R&D efforts would lead to significantly better results for firms with relatively intensive R&D efforts. This assumption will also form an important background for the interpretation of the empirical results obtained in this chapter.

As far as the regional dimension is concerned, we will first consider the empirical results for the generation of *product innovations*. In this context, it can be concluded from Table 8.1 above that the estimated parameters with respect to location - i.e., REG(1) and REG(2) - appeared to be insignificant. Consequently, none of the regions concerned appears to be clearly lagging behind or leading as regards the overall propensity of its firms to generate product innovations.

However, the importance of internal R&D efforts - i.e., RDA(1)- in the generation of product innovations appears to vary across the regions concerned. As can be concluded from the statistically significant interaction effect[8] REG(1) * RDA(1), internal R&D efforts are more important to the generation of product innovations in the central zone than in the peripheral zone. In this context, the intermediate zone holds indeed an 'intermediate' position (see the statistically insignificant interaction effect REG(2) * RDA(1)).

Consequently, the importance of internal R&D efforts - i.e., RDA(1) - for the generation of product innovations in the various regions concerned displays the following spatial pattern:

Figure 8.1   Internal R&D and the Generation of Product Innovations

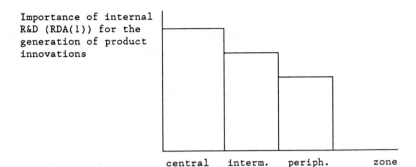

Thus, under the above mentioned assumption concerning the relationship between R&D efforts and quality of innovations, the above pattern points at a rather favourable performance of the central zone - compared to the intermediate and especially the peripheral zone - as regards the *quality* of the product innovations generated.

It has been discussed above that 'new line' firms are more oriented toward the generation of product innovations than 'old line' firms. However, the statistically significant interaction effect BRAN(1) * REG(2) reveals that this sectoral difference is especially important in the intermediate zone while being less important in the peripheral zone. Consequently, it appears that in the intermediate zone considerable differences exist between 'new line' and 'old line' firms as regards the generation of product innovations. In the peripheral zone on the other hand, the 'old line' firms are more able to hold pace with the 'new line' firms as far as the generation of product innovations is concerned. Concerning these sectoral differences the central zone holds an intermediate position (witness the insignificant interaction effect BRAN(1) * REG(1)).

Concerning the spatial differences in the generation of *process innovations* (see Table 8.2 above), the estimated parameter with respect to location in the central zone - i.e., REG(1) - appears to be *negative* (and statistically significant). In this context, the estimated parameter with respect to location in the intermediate zone - i.e., REG(2) - is also negative, though less negative than for the central zone. Consequently, it appears that the central zone is lagging behind

- compared to the intermediate and especially the peripheral[9] zone - in terms of the generation of process innovations.

As far as the generation of process innovations is concerned, the importance of external R&D efforts - i.e., ERD(1) - also appears to vary across the regions concerned. In this context, external R&D efforts appear to be significantly more important in the generation of process innovations in the intermediate zone than in the central zone. This regional discrepancy in the impact of external R&D efforts is reflected by the (statistically significant) variable HULP, which is *positive* for the intermediate zone and *negative* for the central one.

This variable indicates that in the intermediate zone the estimated parameter with respect to external R&D efforts (i.e., ERD(1)) is significantly larger - i.e., with a positive discrepancy of +0.21- than the estimated parameter for the main effect concerning this variable, while the opposite conclusion holds for the central zone. In this context, the peripheral zone holds an intermediate position. The estimated parameter with respect to external efforts in this zone did not appear to differ (significantly) from the main effect concerning this variable.

Consequently, the importance of external R&D efforts - i.e., variable ERD(1) - in the generation of process innovations displays the following spatial pattern:

Figure 8.2 External R&D and the Generation of Process Innovations

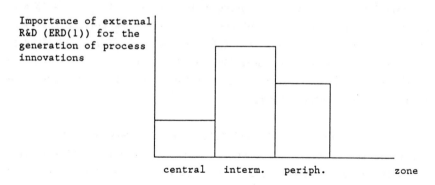

Thus, the above patterns indicate that - besides an overall lower performance with respect to the generation of process innovations - the

central zone also appears to be lagging behind in a qualitative sense compared to the peripheral and (especially) the intermediate zone.

The empirical results presented in this section indicate that regarding the *types* of innovations generated, the central zone has a relatively poor performance with respect to the creation of process innovations but not for the generation of product innovations. These empirical results largely support our theoretical framework - and in particular the fourth hypothesis deduced from this framework - as developed in the first part of this book. In the conclusion to this chapter we will return to this issue.

First however, in the next section we will analyze whether similar patterns of innovative behaviour can also be discerned at the *urban* level of analysis.

8.4 THE URBAN LEVEL OF ANALYSIS

In this section our location variable will be defined at the municipality level. In this context the generation of product and process innovations in various types of Dutch municipalities will be analyzed (see also Davelaar and Nijkamp 1989c, 1989h). As regards the interpretation of the empirical results to be presented in this section we will only consider the *spatial* dimension of innovative behaviour. For a discussion of the impact of the relevant intra-firm variables upon innovative behaviour the reader is referred to the previous section.

In this respect, two different types of classification of Dutch municipalities have been constructed:

(1) a demarcation which is in agreement with the concept of a metropolitan area (see Lambooy 1978, Kok et al. 1985) and which is based on the notion that the sphere of influence of a central municipality goes far beyond its strict (administrative) boundaries. In order to define such metropolitan areas we have used van der Knaap and Sleegers's (1980) analysis, in which central ('nodal') places (so-called A-municipalities) and surrounding places (so-called B-municipalities) which are functionally related

to these central A-municipalities - are regarded as a coherent
urban zone. Consequently, we amalgamated these A- and
B-municipalities as the basis for a broader notion of the
metropolitan milieu.

However, given the average size and location in the economic
centre, A-municipalities located in the central zone[10] (like
Amsterdam, Rotterdam, Utrecht, The Hague) have been dealt with
separately. Consequently, as far as this first regional demarcation
is concerned, the following possibilities will be distinguished
with respect to the location of a firm:

- Location in an A-municipality in the central zone of the
  Netherlands, denoted here as MUN(2)
- Location in an A-municipality elsewhere or a B-municipality,
  denoted here as MUN(1)
- Location in remaining (rural) municipalities denoted here as
  MUN(0)

Thus, municipalities of type 2 - i.e., MUN(2) - may be regarded
as the most urbanized and information-intensive zones in the spatial
hierarchy, whereas municipalities of type 0 - i.e., MUN(0) - belong to
the lowest rank.

(2) a demarcation based on an administrative subdivision according to
     Dutch municipalities; here the innovativeness of firms in large
     and medium size places (i.e., with a central city larger than
     50,000 inhabitants) is compared with the innovativeness of firms
     located elsewhere. These large and medium size municipalities
     belong to the categories C4 and C5 of the Dutch classification
     of municipalities presented by the Central Bureau of Statistics
     (CBS).[11]
     In the context of this second regional demarcation the
     following possibilitiies with respect to location of a firm
     will be distinguished:

- Location in a so-called C4 or C5 municipality, denoted here as
  URB(1)

- Location in other (remaining) municipalities, denoted here as
URB(0)

Given these spatial demarcations, the above mentioned logit
analysis has been performed. Tables 8.3 and 8.4 below contain the
empirical results for regional demarcation (1), while Tables 8.5 and
8.6 present the results for regional demarcation (2).

Table 8.3   Estimation Results with respect to Product Innovations
for Regional Demarcation (1)

Variable	Estimate	Statistical Significance
BRAN(1)	-0.25	-3.5
ERD(1)	0.37	4.4
WERK(1)	-0.14	-2.1
RDA(1)	0.90	6.7
RDA(2)	0.72	6.3
MUN(2)	0.03	-
MUN(1)	0.11	-
ERD(1) * MUN(2)	0.50	} 1%
ERD(1) * MUN(1)	-0.22	
RDA(1) * MUN(2)	-0.38	8%
RDA(1) * MUN(1)	0.10	-

$G^2 = 42.9$, d.o.f. = 60

Table 8.4   Estimation Results with respect to Process Innovations
for Regional Demarcation (1)

Variable	Estimate	Statistical Significance
BRAN(1)	0.10	1.2
ERD(1)	0.35	4.1
WERK(1)	-0.43	-4.7
RDA(1)	0.66	4.8
RDA(2)	0.31	2.6
MUN(2)	-0.11	} 5%
MUN(1)	0.21	
BRAN(1) * WERK(1)	-0.19	2%

$G^2 = 36.1$, d.o.f. = 63

First of all, we will discuss the outcomes for the *first* mentioned regional demarcation of Dutch municipalities developed above. As far as the generation of product innovations is concerned, Table 8.3 above reveals that the estimated parameters with respect to location-i.e., MUN(2) and MUN(1) - appear to be insignificant. Consequently, none of the various types of municipalities distinguished appears to be clearly leading or lagging behind as regards the generation of product innovations.

In this context however, the impact of external and internal R&D efforts - i.e., RDA(1) and ERD(1) - for the generation of product innovations displays a mutually contrasting pattern across the successive types of municipalities distinguished for regional demarcation (1). First of all, external R&D efforts appear to be especially important in the most highly urbanized - i.e., MUN(2)-municipalities (see the positive interaction effect ERD(1) * MUN(2) in Table 8.3). In this respect, the importance of external R&D efforts appears to decline as the urbanization degree declines (witness the negative interaction effect ERD(1) * MUN(1) in Table 8.3).

On the other hand however, the importance of internal R&D efforts - i.e., RDA(1) - for the generation of product innovations displays exactly the opposite pattern. Internal R&D efforts appear to be less important in the highly urbanized - i.e., MUN(2)-municipalities than in the less urbanized - i.e., MUN(0)-municipalities (see the negative interaction effect RDA(1) * MUN(2) in Table 8.3 above). In this respect, the intermediate urbanized - i.e., MUN(1) - municipalities hold an 'intermediate' position (see the insignificant interaction effect RDA(1) * MUN(1) in Table 8.3).

Consequently, on the basis of the first mentioned regional demarcation of Dutch municipalities the following spatial pattern emerges concerning the importance of internal and external R&D efforts for the generation of product innovations (see also Davelaar and Nijkamp 1987c):

Figure 8.3 Urbanization Degree versus Internal and External R&D

Importance of internal
and external R&D efforts
for the generation of
product innovations

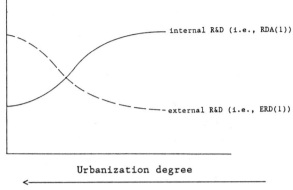

internal R&D (i.e., RDA(l))

external R&D (i.e., ERD(l))

Urbanization degree

When the urbanization degree is rising, the importance of external R&D for product innovations rises, while then the importance of internal R&D seems to decline. Which factors might explain the above mentioned spatial patterns? In this context three possibilities may be suggested here (see also Davelaar and Nijkamp 1987c):

(1) There is a 'spatial substitution mechanism' between internal and external R&D. Firms located close to nodes of an information network (i.e., in a metropolitan milieu), will very much benefit from external R&D contacts and hence are less dependent on internal efforts.

(2) Various types of industrial activity are strongly dependent on external R&D for their product innovations and are henceforth 'forced' to locate near information or incubation nodes. Other types of firms are able to compensate for a less central location by means of an increase of internal R&D. Thus (functional) differences between firms are relevant here (cf., Ewers 1986).

(3) Internationally oriented firms are biased toward the central metropolitan areas (cf., Oakey et al., 1980) from which they maintain their internationally (external) R&D contacts.

As far as the generation of *process innovations* is concerned, spatial discrepancies appear to be more important than for the creation of product innovations. As Table 8.4 above reveals, the estimated parameters with respect to location - i.e., MUN(2) and MUN(1) - indicate that firms located in the intermediate urbanized (MUN(1)) municipalities have a significantly leading position regarding the generation of process innovations compared to firms located in the other (MUN(0) and MUN(2)) municipalities.

Consequently, our empirical results for regional demarcation (1) indicate that as far as the creation of product innovations is concerned, firms located in the most urbanized (MUN(2)) municipalities do *not* lag behind compared to firms located elsewhere. Concerning the generation of process innovations however, these highly urbanized areas appear to be lagging behind in comparison to the intermediate urbanized (MUN(1)) municipalities.[12]

These results appear to be largely consistent with the results found for the *regional* level of analysis as discussed in the previous section. Also for the regional level of analysis the most urbanized-i.e., central - zone of the Netherlands appears to perform relatively poorly concerning the generation of process innovations but not concerning the creation of product innovations.

Table 8.5   Estimation Results with respect to Product Innovations
for Regional Demarcation (2)

Variable	Estimate	Statistical Significance[13]
BRAN(1)	-0.25	-3.6
ERD(1)	0.26	3.6
WERK(1)	-0.14	-2.0
RDA(1)	0.92	7.2
RDA(2)	0.72	6.4
URB(1)	0.12	1.5
URB(1) * ERD(1)	0.16	} 3%
URB(1) * RDA(1)	-0.26	

$G^2 = 37$, d.o.f. $= 39$

Table 8.6   Estimation Results with respect to Process Innovations
            for Regional Demarcation (2)

Variable	Estimate	Statistical Significance
BRAN(1)	0.10	1.2
ERD(1)	0.28	3.1
WERK(1)	-0.44	-4.8
RDA(1)	0.67	4.6
RDA(2)	0.34	2.6
URB(1)	-0.13	-1.4
BRAN(1) * WERK(1)	-0.20	2%
URB(1) * ERD(1)	-0.17	8%

$G^2 = 34$, d.o.f. = 39

Finally, our empirical results for regional demarcation (2) - as summarized in Tables 8.5 and 8.6 above - will briefly be considered here.

Although not significant at the 10% level of statistical significance, the estimated parameters for the variable 'location' in both tables correspond with the spatial innovation patterns discussed above. In this respect, the estimated parameter with respect to location in 'urban' - i.e., URB(1) - municipalities is (clearly) *positive* for the generation of *product* innovations, while being (clearly) *negative* for the creation of process innovations.

As far as the generation of *process innovations* is concerned, external R&D efforts appear to be less important in the 'urban' - i.e., URB(1) - municipalities than in the 'non-urban' - i.e., URB(0)- municipalities (witness the negative interaction effect URB(1) * ERD(1) in Table 8.6 above). This pattern indicates a relatively poor performance of the 'urban' municipalities as regards the *quality* of the process innovations generated (see the previous section) and corresponds to the results found for the *regional level* of analysis (see Figure 8.2 of the previous section).

Also in line with our prior results found for regional demarcation (1) is the spatial differential impact of internal and external R&D - i.e., RDA(1) and ERD(1) - efforts upon the generation of *product innovations.* The empirical results found for regional demarcation (2) also indicate that external R&D efforts are more important in the 'urban' - i.e., URB(1) - than in the 'non-urban'- i.e., URB(0) - municipalities (witness the positive interaction effect URB(1) * ERD(1) in Table 8.5). Thus (again) these results seem to imply that in nodal points of a network which are marked by high information intensities (see Davelaar and Nijkamp 1987c) - i.e., the URB(1) municipalities - external R&D is more important to the generation of product innovations than elsewhere. In this context however, also for this second regional demarcation internal R&D efforts appear to be relatively more important in the 'non-urban' - i.e., URB(0)- municipalities (see the negative interaction effect RDA(1) * URB(1) in Table 8.5 above).

In conclusion, also for regional demarcation (2) the importance of internal and external R&D efforts for the generation of product innovations appears to support the spatial patterns depicted in Figure 8.3 above.

8.5 CONCLUSION

In this chapter we have particularly concentrated upon the generation of both product and process innovations in Dutch manufacturing firms. By using a multivariate logit analysis, the impact of various intra-firm variables upon the generation of both types of innovations has been analyzed. In this respect, also spatial differences - both at the regional and urban level of analysis - have been considered.

As far as the impact of the intra-firm variables is concerned, our analysis has demonstrated that 'new line' firms are more oriented toward the generation of product innovations than 'old line' firms. On the other hand however, especially the large 'old line' firms appear to be forerunners in the generation of process innovations. These results largely confirm the innovation life cycle concept discussed in the first part of this book. According to this concept the creation of (further) product innovations will be especially important in earlier phases of the life cycles of industries and (underlying) technological 'trajectories'. In later phases however, the main emphasis of the innovative activities will become increasingly oriented towards further improving process technologies.

Firm size also appears to be important in the capability of a firm to generate product and process innovations. However, our results also suggest that firm size is particularly important in the generation of process innovations.

In this context, internal and external R&D efforts are also important in generating both product and process innovations. However, internal R&D efforts appear to be more important for the generation of product innovations than for the generation of process innovations.

As far as the spatial dimension of innovative behaviour is concerned, our results for the *regional* level of analysis indicate that the central zone of the Netherlands performs relatively poorly-compared to the intermediate and the peripheral zone - as regards the generation of (high quality) *process* innovations. On the other hand however, this conclusion does not hold for the generation of product innovations. Concerning these innovations our results even point at a relatively favourable performance of the central zone concerning the *quality* of the product innovations generated.

It is noteworthy that rather similar innovation patterns could be disentangled for the *urban* level of analysis. For this level of analysis, two different regional demarcations of Dutch municipalities have been considered. Our results indicate that the most urbanized municipalities perform relatively poorly regarding the generation of (high quality) process innovations but not in terms of the generation of product innovations.

As far as the urban level of analysis is concerned, our analysis also suggests that the importance of internal and external R&D efforts for the generation of product innovations varies - and in an opposite pattern - according to urbanization degree. In this respect, external R&D efforts appear to be especially important in the most urbanized areas, while firms located in the less urbanized areas appear to be more dependent upon their own internal R&D capabilities.

This chapter concludes our analysis of the spatial innovation patterns for the Dutch manufacturing sector. In recent decades, the employment generating capacity of this sector has declined rather drastically. In the first part of this book, it has been discussed that the decline of (employment in) the manufacturing sector can (partly) be ascribed to a relative lack of basic innovations and resulting new technological 'trajectories' (cf., Freeman et al 1982, van Duijn 1983, Rothwell and Zegveld 1985).

As far as the *spatial* dimension is concerned however, especially the most central (metropolitan) regions appear to have suffered from this decline of the manufacturing sector (see also Chapter 5 of this study). In various explanations provided for this spatial shift in manufacturing employment especially the role of a-technological factors has been stressed. In our dynamic incubation framework developed in the first part of this book however, we also called attention to the role of innovative activities in generating such a spatial shift in later phases of the life cycles of industries and technologies.

In Chapter 5 we found that (several regions located in) the central zone of the Netherlands had a relatively poor performance from the viewpoint of employment. Further, the empirical results in Chapters 6 and 7 suggest that the innovation potential in several of these central (metropolitan) regions is relatively weak for various types of manufacturing firms. As explained in these chapters, innovation potential is mainly concerned with innovation from the *input* side (e.g., R&D efforts). At the *output* side however, the spatial patterns appear to differ for product and process innovations. The results of the present chapter suggest that regarding the *types* of innovations generated, the relatively poor performance of these central regions applies especially to the generation of *process innovations* but not to the generation of product innovations.

These empirical findings largely support our theoretical framework - and in particular the third and fourth hypothesis deduced from this framework - as developed in Chapter 3. These hypotheses state that in later phases of the life cycles of industries and technologies non-metropolitan or non-central regions may become dominant concerning the innovative activities performed (i.e., hypothesis 3: see the results of the previous chapters). As to the types of innovations generated in these later phases however, this 'dominance' will especially apply to the generation of process innovations (i.e., hypothesis 4: see the results of the present chapter).

In the next chapter, the second hypothesis deduced from this framework (i.e., the hypothesis that new 'Schumpeterian' industries will proceed from metropolitan or central regions onwards) will be tested empirically. In that context the spatial development pattern of several (new and innovative) producer service sectors will be analyzed.

Notes to Chapter 8:

1. The term 'new' refers here to the opinion of the firms interviewed (see Chapter 6).

2. Details on logit analysis can be found in Bishop et al (1977) and Nijkamp et al (1985).

3. So these are firms claiming to have introduced one or more product or process innovation(s) respectively in 1983 which were - according to the firm - new to the whole branch of industry in the Netherlands in which the firm operates.

4. Also for these interaction effects the statistical level of significance has been determined in a similar way as described for the main effects with respect to the variable 'location'.

5. For an exact definition of these zones the reader is referred to Chapter 7 of this book.

6. As regards the estimated parameters for the main effects with respect to the intra-firm variables the ratio between the estimated parameters and its standard error has been reported in all tables.
As stated before, this ratio is comparable to t-values in regression analysis (and provides information about the statistical level of significance of these parameters).
As stated in the previous section, in order to determine the statistical significance of the main effects with respect to location-i.e., REG(1) and REG(2) - and the higher order interaction effects we compared the improvement in the likelihood ratio G^2 with the change (i.e., loss) in the total number of degrees of freedom resulting from including these extra parameters. Brackets indicate that (regional) main or interaction effects have been added simultaneously to the estimated model and the statistical significance refers to the improved fit of the estimated model resulting from such a combined addition. Insignificant results are indicated by a - sign in the tables.

7. Given the above mentioned restrictions upon the estimated parameters, this interaction effect is positive for the large 'old line' and small 'new line' firms, whilst it is negative for the small 'old line' and large 'new line' firms.

8. Recall again that this interaction effect is negative for the peripheral zone; see the restrictions specified in the previous section with respect to the estimated parameters.

9. Recall again that the estimated parameter with respect to location in the peripheral zone - i.e., REG(3) - is equal to +0.21 (i.e., the negative sum of the estimated parameters with respect to location in the central and intermediate zone).

10. For a definition of the central zone of the Netherlands: see the previous chapter.

11. In the previous chapters these municipalities have been denoted as DUM1 areas.

12. Similar conclusions hold for the performance of the less urbanized-
i.e., MUN(0) - municipalities.

13. In Tables 8.5 and 8.6 also for the main effects with respect to the
variable 'location' the estimated parameters divided by their standard
errors have been recorded.

9 Spatial dispersion of producer services in the Netherlands

9.1 INTRODUCTION

In recent years the role of information in economic production processes has increased considerably (cf., Naisbitt 1984, Noyelle et al 1984, Huppes 1987, de Smidt et al 1987, 1989, Orishimo et al 1988, Giaoutzi et al 1988, Davelaar and Nijkamp 1989a, 1989b). In this context, basic innovations in the field of computer and telecommunication technology - in interaction with socio-institutional changes (innovations) - are at the heart of the more recent information 'technology system' (cf., Freeman et al 1982, Freeman 1987a, 1987b; see section 9.2 of this chapter). Regarding the prevailing technological paradigm of the (new) technological 'trajectories' one can observe a shift from an energy-intensive mass-production orientation to an information intensive orientation (cf., Freeman 1987b; see also Chapter 1).

In this respect, (new) sectors related to the above-mentioned information 'technology system' are rapidly expanding nowadays. It is noteworthy here, that in particular the producer service sector-providing in particular various types of information services - is often considered to be such a dynamic growth sector (cf., Pred 1977, Bearse 1978, Lambooy et al 1983, Noyelle et al 1984, Marshall 1985, Van Dinteren et al 1988 and Johansson et al 1989). Pred, for example, remarks:

In recent years, rapid market and technological changes have brought about a strongly increased demand for business services. Changes in the economic and technological environment have also dictated that these business services become ever more specialized (Pred 1977, pp.119).

In this context, de Haan and Tordoir (1986) have shown that - as far as the Dutch context is concerned - especially in 'high tech' sectors which exhibit high values added (like electronics, instruments and producer services itself) the role of information inputs in the production process has increased considerably in recent years. This is reflected in both a growing share of employees *within* these sectors being engaged in knowledge intensive functions and in a growing input of *external* producer services. Consequently, the above patterns point at a 'structural adjustment process' in which information tasks become increasingly important both within firms and between firms mutually.

The growing importance of producer service activities has also resulted in a 'Schumpeterian swarming process' of (new) entrepreneurs taking up the new market opportunities opened up by the more recent 'information technology system'. In this context, Alderman et al (1986, pp.8), for example, conclude that 'a lot of the opportunities for new firms are occuring in new service activities, resulting from technological advancement, the best example perhaps being software services'. In this respect, also the importance of producer services for local innovative activities is increasingly being stressed (cf., Oakey et al 1980, Ewers 1986, Bade 1986).

Consequently, the producer service sector may not only be relevant from a national-economic point of view, but also from a meso-spatial point of view as it may offer opportunities for restructuring metropolitan economies which in the past decades have been under severe pressure as far as employment in the manufacturing sector is concerned (see Chapter 5). Given these considerations, in this chapter we will particularly concentrate on the spatial distribution of selected producer service (sub)sectors in the Netherlands.

It is noteworthy that this research issue is also important for an empirical test of our theoretical framework. The second hypothesis deduced from this framework suggests a 'hierarchical' spatial diffusion pattern of such new 'Schumpeterian' sectors related to the more recent 'information technology system'.

This chapter is organized as follows. In section 9.2 the increased importance of the producer service sector as a whole for the Dutch economy will briefly be considered. Section 9.3 will be devoted to a selection of special subclasses of the producer service sector which may be considered to be highly dynamic and largely affected by the above mentioned growing importance of (the handling of) information in the economic system. Section 9.4 will contain a discussion of the characteristics of our empirical data base and of some general empirical findings. On the basis of these findings we will also make a final selection of producer service subclasses which will be analyzed in a spatial context. Next, in section 9.5 the spatial dispersion of the selected producer service subsectors will be analyzed more in-depth. Section 9.6 will be devoted to a further consideration of the 'time-space trajectory' of the selected subclasses. Section 9.7 contains some general conclusions.

9.2 PRODUCER SERVICES IN THE NETHERLANDS

The aim of this section is to provide some empirical evidence on the growing economic importance of the producer service sector in the Netherlands. We will mainly confine our analysis to those firms which belong to SIC-code 84 (notably computer service, marketing offices, management consultancy to mention a few). Thus we will exclude banking and insurance sectors which also supply several intermediate services. The reason for doing so is that we consider these sectors to be relative 'old' sectors - i.e., compared to SIC-sector 84 - which are not dominated by 'entrepreneurial innovation' and a resulting high level of new firm formation. As it is our aim to analyze the spatial dispersion of innovative and dynamic sectors - in the sense of new 'Schumpeterian' entrances - we have excluded these sectors à priori from our analysis.

Clearly, any attempt to give an in-depth statistical analysis of the historical growth pattern of the producer service sector is very difficult. Data for this sector are relatively scarce compared to 'old' sectors like agriculture and manufacturing (cf., Lambooy and Tates, 1983). Also, statistical revisions and re-definitions hamper a profound time series analysis of this sector regarding such variables as employment, value added and production. These problems become even more troublesome, if - like in our case - we want to study the development within several sub-components of the producer service sector.

The growing economic importance of the producer service sector is best illustrated by means of employment data - i.e., number of jobs- which we derived from the Statistics of Employed Persons.[1] As can be seen below, these data can - for various years - be obtained at a 3-digit level. Our 3-digit figures relate mainly to the seventies as after 1979 employment data at a 3-digit level are only presented for firms employing more than 10 employees.

Table 9.1 Jobs in the Producer Service Sector in the Netherlands

SECTOR \ YEAR	1973	1975	1977	1979	1981	1983	1985
Legal services (841)	7,876	9,089	10,390	10,831			12,685
Accountancy/tax advice (842)	29,375	30,250	37,109	45,880			63,013
Computer services (843)	61	1,217	2,824	4,723			13,615
Engineering (844)	33,851	35,053	41,912	48,440			51,637
Advertising (845)	6,565	7,295	6,640	8,782			10,197
Consultancy (846)	7,096	7,531	7,981	9,030			9,964
Press agencies (847)	569	588	412	598			661
Secretarial agencies (848)	27,852	37,671	38,732	47,086	38,500		102,430
Remaining categories (849)	22,611	21,347	20,222	23,556			28,219
Total number of jobs	3841,241	3914,055	3966,337	4117,403	4161108	3976735	4480,510
Share of jobs in SIC 84	3.54	3.83	4.19	4.83	4.96	5.18	6.53

Source: Statistics of
Employed Persons
(several years)

It should be noted that before 1984 only jobs of 15 working hours a week or more have been counted in these statistics, while, from 1984 onward, also jobs with less than 15 hours a week have been included.[2] This explains the rather drastic increase in the total number of jobs between 1983 and 1985 depicted in this table.

The growth of employment (i.e., jobs) in the period 1973-1985 applies to various (3-digit) sub-groups of the producer service sector. However, especially the growth of the computer service sector has been explosive. On the other hand, growth of employment in secretarial agencies has been of a fluctuating nature. These fluctuations might partly be a reflection of the more general economic tide on which this sector seems to be dependent.

Also the *share* of jobs in the producer service sector - compared to the *total* number of jobs - has steadily been increasing (see the last row of Table 9.1). Thus, Table 9.1 quite clearly demonstrates the growing importance of various producer service sub-sectors as a pool of new jobs. It is noteworthy that the number of jobs in the manufacturing sector as a whole has steadily been decreasing during this period in the Netherlands. Figure 9.1 below summarizes the development of jobs in manufacturing compared to the producer service sector (i.e., SIC 84).

Figure 9.1   Jobs in the Producer Service and Manufacturing Sector

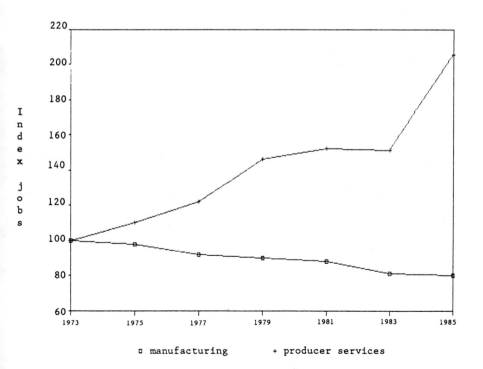

□ manufacturing   + producer services

Figure 9.1 clearly depicts two opposite patterns. On the one hand the steady rise of jobs in the producer service sector is clearly reflected in this figure. On the other hand, the manufacturing sector is rapidly losing its role as a job-provider. Even the statistical re-definition of 1984 (i.e., also including jobs with less than 15 hours a week) did not alter the rather poor 'performance' of the manufacturing sector during the period 1973-1985.

The increased importance of the producer service sector in total intermediate consumption of firms and in total net value added of firms is reflected in Tables 9.2 and 9.3, respectively. In Table 9.2 time-series have been constructed that are more or less comparable (denoted as 'comparable' years).[3]

Table 9.2   Intermediate Deliveries provided by the Producer Service
            Sector

'Comparable' years	1972	1973	1976	1977
(1) total intermediate consumption of firms (mln.Dfl)	78,390	90,390	130,300	138,780
(2) amount of (1) provided by SIC 84/85 (mln.Dfl)	1,440	2,170	3,510	3,730
(3) share provided by SIC 84/85	1.83	2.40	2.69	2.69

'Comparable' years	1977	1978	1979	1980	1981	1982	1983
(1) total intermediate consumption of firms (mln.Dfl)	133,000	140,000	153,000	167,000	178,000	198,000	205,000
(2) amount of (1) provided by SIC 84/85 (mln.Dfl)	6,480	7,490	8,330	9,290	9,810	10,378	11,067
(3) share provided by SIC 84/85	4.87	5.35	5.44	5.56	5.51	5.24	5.40

Source: National Accounts
(several years)

Table 9.3 Net Value Added (Market Prices) of SIC 84/85

year	net value added SIC 84/85 (mln.Dfl)	total net value added of firms (mln.Dfl)	share SIC 84/85 in total net value added
1971	4,145	108,780	3.81
1972	4,582	122,610	3.74
1973	5,336	140,230	3.81
1974	6,340	157,640	4.02
1975	7,345	170,920	4.30
1976	8,830	196,640	4.49
1977	10,501	214,480	4.90
1978	12,170	230,900	5.27
1979	13,280	244,390	5.43
1980	14,250	259,800	5.48
1981	14,600	278,850	5.37
1982	14,948	283,900	5.27
1983	16,112	294,480	5.47

Source: National Accounts 1969-1981
(revisions for the years 1969-1976)
National Accounts 1985
Depreciation figures for the SIC
categories have been derived from the
original National Accounts 1973-1985

In interpreting these tables, however, one should be aware of the fact that the statistical 'ignorance' effect of the producer service sector is also reflected in these tables, as the statistics are not available at a 2-digit (SIC 84) level. Consequently, the data relate to an aggregation of SIC 84 and 85 (rental companies).[4]

Both tables show that the general pattern mirrors a growing economic importance of the producer service sector, in terms of their share in both intermediate deliveries and in total net value added. The increasing performance of this sector seems to have slightly stagnated in 1981 and 1982, but, as Table 9.1 reveals, this might to a large extent be due to the rather drastic decline of SIC 848 (secretarial agencies) in these years. However, it is not possible to fully test this assumption given the aggregation level of the available data.

In 1983 however, both tables show again a 'recovery' of the producer service sector. So the hypothesis that producer services 'perform an increasingly pivotal role as intermediaries in the production of both goods and services' (Daniels 1987, pp.1) is indeed confirmed by our data.

In the next section we will select special subclasses of the producer service sector which may be considered to be highly dynamic and largely affected by the more recently evolving information 'technology system'.

9.3 SELECTED PRODUCER SERVICE GROUPS

In the foregoing section we noticed the scarcity of data on the growth of the producer service sector which is reflected inter alia in the number of jobs, net value added and intermediate deliveries. At the *spatial* level however, these data limitations become even more severe. Besides, even at the 2-digit level (for which some data are available) the producer service sector consists of rather heterogeneous subclasses. Consequently, the spatial dispersion of the (innovative subgroups of the) producer service sector is difficult to trace due to a serious lack of available data.

Therefore we had to construct our own data set for the Netherlands with the help of information from the Chambers of Commerce in order to carry out a more disaggregate, spatial (and dynamic) analysis. For this purpose we have first selected special subclasses of the producer service sector which may be considered to be largely affected by the more recent 'information technology system'. The framework for selecting these subgroups is sketched in Figure 9.2 below.

Figure 9.2   Producer Services and the Information 'Technology System'

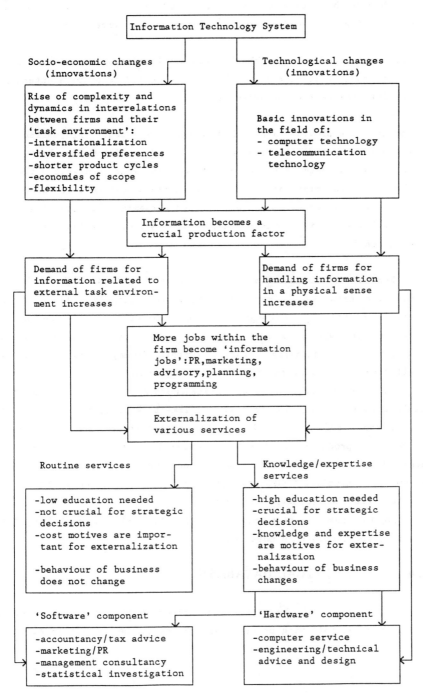

Thus we have used two (interrelated) criteria for selecting and distinguishing two broad categories of producer service firms:

1. The services provided by these firms are knowledge-intensive, i.e., those producer service firms which are expected to provide external services for reasons of expertise.[5]

2. The services provided are linked to the more recent 'information technology system'. In this respect, we made a distinction between services that cater for the need of mere information itself - i.e., the 'software component' - and those services that meet the need of the mere capacity to store, process and disseminate information (flows) itself within firms, institutions - i.e., the 'hardware component'.

Recent developments with respect to CAD/CAM, CNC machines, information networks to mention a few, have largely influenced the services provided by the engineering and computer service sector. As these services are to a high degree oriented towards increasing the mere information handling *capacity* (within firms, institutions), we labelled these services the 'hardware' component.

The general increased need of information itself - resulting from highly dynamic and complex changes in firm-'task environment' relations (cf., Lambooy and Tordoir 1985) - has also been an important impulse for various producer service firms engaged in these relations. As these firms are not principally concerned with increasing the information handling *capacity* (of firms, institutions) they are labelled here the 'software' component.

In the next section the characteristics of our empirical data base will be discussed in more detail.

9.4 THE EMPIRICAL DATABASE

On the basis of the criteria described in the foregoing section, we selected the following provisional sectors at the 4-digit level:

'Software' subgroup:

Financial services:
SIC-code:	Sector:
8421	Accountancy
8423	Tax advice

Non-financial services:
SIC-code:	Sector:
8451	Advertising
8461	Market analysis
8462	PR-advice
8464	Statistical analysis
8466	Management consultancy

For reasons to be indicated later on, regarding the 'software' component we made a distinction between firms into those mainly performing *financial* and *non-financial* services. Firms providing these types of services will also be denoted as the 'general information' subgroup.

'Hardware' subgroup:

Computer services:
SIC-code:	Sector:
8431	Computer service

Engineering advice:
8441	Engineering/technical advice and design
8443	Construction advice and design
8449	Remaining engineering advice/design

As stated before, the data situation for the producer service sector -especially at the 3-digit and 4-digit levels - is rather poor in the Netherlands, especially if one wants to study space-time patterns. For reasons of confidentiality the Central Bureau of Statistics does not publish such data at a very disaggregate (regional) level.

Consequently, in order to circumvent these data problems we had to construct our own database. In this respect, the data collected by the Dutch Chambers of Commerce (36 in total) (and centrally available at the Central Databank in Woerden) appears to be the best substitute.[6] In the Netherlands, (nearly) every establishment, whether main or subsidiary establishment, has the legal obligation to register itself to the pertaining Chamber of Commerce.

Consequently, these Chambers do not only possess data on the number of establishments, but also on the number of employees, main activities of the establishment (i.e., SIC-codes), location and the like.[7] The Central Databank does not only register data from the regional Chambers, but employs also other sources (e.g., inquiries) in updating its database.

Consequently, we decided to use this database with respect to our selected producer service sub-groups. In order to minimize costs, we constructed the following sample design with respect to size categories and sectors (see Table 9.4).

Table 9.4 Sample Design for Selected Producer Service Subgroups

number of employees	probability of being selected	total number of establish- ments	financial information group 1	computer service group 2	engineering advice/design group 3	general information group 4
1	0.1	11,704	1,019 (111)	2,352 (263)	2,805 (297)	5,528 (625)
2-4	0.333	4,179	365 (125)	795 (290)	995 (328)	2,024 (713)
>4	1.0	3,072	427 (459)	666 (751)	1,055 (1,136)	924 (1,004)

Figures without parentheses refer to the situation of May 1,1987 - representing total number of establishments - when the Central Databank made several cross-section counts. Figures in parentheses are based on our own *sample* results which represents the *December 1987* situation. Data have been gathered at the *establishment* level.[8] As can be derived from Table 9.4 above, only small establishments - i.e., smaller than 5 employees - have not been integrally observed in our database.

It is evident that the number of establishments (and the related employment in these sectors) appears to have increased considerably even in the relative short period May-December 1987. This applies especially to the *computer service* and '*general information*' group. In these cases the number of establishments with more than 4 employees - which were integrally observed in our sample - increased in this period. Also the observed number of establishments in the size categories 1 and 2-4 employees - for which we developed a stratified sample design; see Table 9.4 above - clearly exceeds the number that might have been expected on the basis of the (total) May 1987 figures presented by the Central Databank.[9]

On the other hand, in the financial and engineering advice subgroups (i.e., group 1 and 3 in Table 9.4 above) development in the total number of establishments appears to be less expansive, though still growing. In this case the number of establishments larger than 4 employees has increased in both subgroups. Also the number of establishments with 1 employee seems to have increased in both sectors in this period (though not in a statistically significant way: see note 8). In the size category 2-4 employees however, the observed sample outcomes largely correspond to the 'expected' outcomes on the basis of the May 1987 figures.

Consequently, our expectation that the selected subgroups can be considered as highly dynamic - i.e., many new 'Schumpeterian entrances' - is indeed confirmed by the data. These signs of highly dynamic subsectors are also reflected in Table 9.5 below in which the estimated number and percentage of 'new' establishments - i.e., those establishments that have a date of establishment after December 31, 1984 - have been presented.[10]

In the first row of this table the estimated total number of establishments is presented. In the second row the estimated total number of establishments larger than 1 employee has been presented. The third row contains the estimated total number of 'new' establishments - i.e., with a 'date of establishment' after December 31, 1984.[11] Row 4 contains the estimated total number of 'new' establishments larger than one employee. Next, row 5 contains the (estimated) percentage of 'new' establishments, while row 6 contains the same percentage for establishments larger than 1 employee.[12]

Table 9.5 'New' Establishments in Selected Producer Service Subgroups

	Financial Information	Computer Service	Engineering Advice/Design	General Information
estimated total number of establishments Dec. 1987	1,944	4,251	5,090	9,393
estimated total number of establishments > 1 employee	834	1,621	2,120	3,142
estimated total number of 'new' establisments	489	2,300	1,710	3,983
estimated total number of 'new' establ. > 1 employee	79	710	460	913
percentage of 'new' establishments (row 3/row 1)	0.25	0.54	0.34	0.42
percentage of 'new large' establ. (row 4/row 2)	0.09	0.44	0.22	0.29

The results in this table (again) indicate that the computer service, 'general information' and 'engineering' subgroups are - and in this order - more 'dynamic' sectors than the financial subgroup (and architects).[13] Especially the first mentioned three subgroups appear to be expanding rather rapidly and characterized by high 'Schumpeterian' entrance rates.

Even when considering only ('new') establishments larger than 1 employee, the results presented in the last row of Table 9.5 stress the importance (and successful nature) of new entrances in these 3 subgroups. In this case the estimated share of establishments younger than 3 years amounts to 44%, 29% and 22% for the computer service, 'general information' and engineering subgroups, respectively. The figures for the financial information subgroup on the other hand, display a far more 'static' pattern.

So we can conclude that especially the computer service, 'general information' and engineering subclasses (and in this order) can be considered as highly dynamic and relatively young sectors performing various (new) information services which are becoming increasingly important in the economic system. So, these subgroups can indeed be provided with the label 'entrepreneurial innovation' (cf., Bearse, 1978). Consequently, in the subsequent sections we will concentrate on the spatial dispersion of (employment in) these three subgroups.

9.5 SPATIAL DISPERSION OF EMPLOYMENT IN SELECTED PRODUCER SERVICE SUBGROUPS

In the present section we will especially concentrate on the spatial dispersion of the computer service, 'general information' and engineering subgroup. On the basis of our foregoing analysis the labels 'new' (i.e., many entrances) and 'innovative' (i.e., linked to the more recent 'information technology system') appear to be best suited for these subgroups. It is clear however, that a certain fraction of these establishments - especially in the engineering and 'general information subgroup - might still be rather old. In this context, it should be noted that at the sectoral level the notions 'new' and 'innovative' are relative concepts (i.e., compared to other sectors).

In order to analyze the spatial dispersion of these selected producer service subgroups the following types of regions will be distinguished:

1. A 'metropolitan' area, which consists of the following COROP regions: the Greater Amsterdam Area, the Greater Rijnmond Area, the Metropolitan Region The Hague, Utrecht and Arnhem/Nijmegen.[14]

2. A second group - denoted here as the 'overspill' area -, which consists of COROP regions surrounding the 'metropolitan' area mentioned above. This 'overspill' area - largely consisting of COROP regions located in the central and intermediate zone of the Netherlands - contains the following COROP regions: Veluwe, Achterhoek,[15] Zuidwest-Gelderland, Kop van Noord-Holland, Alkmaar, IJmond, Agglomeratie Haarlem, Zaanstreek, Gooi en Vechtstreek, Leiden en Bollenstreek, Delft en Westland, Oost Zuid-Holland, Zuidoost Zuid-Holland and ZIJP.

3. The 'performance' of the *province of Noord-Brabant* will be
considered separately, in view of the rather favourable performance
of the Brabant area concerning the (development of employment in
the) manufacturing sector (see Chapter 5, for example).
Thus the Brabant area includes all COROP regions located in the
province of Noord-Brabant.

4. A fourth group - denoted here as the *southern peripheral
areas* - which consists of COROP regions located in the provinces
of Zeeland and Zuid-Limburg.

5. A final group - denoted here as the *northern peripheral areas* -,
which consists of COROP regions located in the northern provinces
of the Netherlands (i.e., Groningen, Friesland, Drenthe and
Overijssel).

It is evident that the first area distinguished above is largely
'*metropolitan*'; the second and third areas are *(semi-)intermediate*,
while the fourth and fifth group are largely *peripheral*.

Our empirical results are based on 'estimated' employment data
concerning the three producer subclasses distinguished. These
estimations are based on our sample derived from the Chambers of
Commerce described in the foregoing section. Consequently, the
estimations provided in this section represent the situation at *the end
of 1987.*[16]

As the Central Databank is not allowed to report the exact number
of employees within the establishments, the following (employment) size
categories are used: 1, 2-4, 5-9, 10-19,20-49, 50-99, 100-199, 200-499,
500-749, 750-999, and >999. Therefore, we used the median of these
categories in estimating our employment figures.[17]
Table 9.6 below contains the (estimated) regional shares of employment
for each of the three producer subclasses selected.

Table 9.6 Estimated Regional Shares of Employment within Selected Producer Service Subgroups

region	regional share of employment in services (SIC 6,7,8) March 1986*	idem manu- facturing*	regional share of estimated employment in computer serv. Dec. 1987	idem 'general information'	idem 'engi- neering'	% of employ- ment within region in subsidiary establishm.	% of popu- lation in 1983**
Metropolitan	44.8	24.7	54.9	54.8	51.1	10.0	31.3
'Overspill'	22.3	24.4	23.1	20.6	19.8	7.5	25.3
Brabant	12.2	20.8	11.7	13.3	9.9	18.2	14.6
Southern Peripheral	7.7	12.1	3.1	3.5	5.0	17.2	10.0
Northern Peripheral	13.0	17.9	5.8	7.7	14.0	23.5	18.3
(Estimated) total employ- ment	1538,000	941,700	28,749	31,571	42,452		

Source: * Statistics of Employed Persons 1986
** Regionaal Statistisch Zakboek 1984, pp.229

In column 2 of this table the regional shares in (total) commercial service jobs (i.e., for the 1-digit SIC categories 6,7,8) has also been given. In the third column of this table the same has been done for the manufacturing sector.[18]

A first conclusion that can be derived from this table is the metropolitan orientation of the commercial service sector as a whole compared to the manufacturing sector. This 'metropolitan bias' however, applies even more so for the selected highly dynamic (and innovative) subclasses of the producer service sector. In this context, the five 'metropolitan' COROP regions as a whole are estimated to possess more than 50% of employment in each of the three producer service subgroups distinguished. Consequently, the 'performance' of these metropolitan areas with respect to employment in these subgroups appears to be (far) better than their 'performance' concerning the total commercial service and (especially) the manufacturing sector.

On the other hand, both the *northern* and *southern peripheral areas* exhibit exactly the opposite pattern. They attract a considerable smaller fraction of total commercial service employment compared to their manufacturing share. However, their 'performance' concerning the selected dynamic producer service subgroups appears to be even more poor. While attracting (together) about 30% of manufacturing employment, their share of employment in the computer service and 'general information' sector is estimated to be only about 10%. The only exception to the rather poor 'performance' of the peripheral areas concerning the selected subclasses appears to be the 'score' of the northern peripheral area with respect to the engineering subgroup.[19]

The *'overspill' region* appears to be the most 'homogeneous' area in terms of employment shares of the various sectors distinguished in Table 9.6: all (estimated) shares can be located in a rough interval of approximately 20-24%. The *Brabant region* finally, in general appears to attract a 'normal' - i.e., compared to its share in total commercial service employment - fraction of employment in the selected subclasses. However, these fractions are considerable smaller than their manufacturing share.

In column 7 of Table 9.6, we have tried to gain some further insights into the question whether a regional bias as regards the 'branch plant' component can be identified. For this purpose, we have estimated the fraction of total employment in the selected subclasses-within each area distinguished - in subsidiary establishments. As only about 10% of the total number of establishments in the sample could be designated as subsidiary establishments, we estimated an aggregate-i.e., for all three subclasses together - figure in this column.

From column 7 of this table it appears that the 'branch plant' component exhibits a regional bias. This effect more or less displays a centre-periphery pattern. The fraction of regional employment in subsidiary establishments appears to be relatively large in the Brabant, southern and (especially) the northern peripheral areas. As can be derived from Table 9.6 above, this fraction is considerable smaller in the 'metropolitan' and 'overspill' areas.

Consequently, these results suggest that in the more peripheral areas the selected producer service subclasses are not only under-represented, but also depend more upon external initiatives than in the more central regions. However, the results in column 7 of Table 9.6 also indicate that only a minor fraction of total employment in the selected producer service subclasses within these former regions is of the 'branch plant' type (also stressing the importance of an indigenous component in these areas).

So far we have presented a rather static description of the spatial dispersion of the producer service subclasses. In the next section the space-time dimension will be dealt with.

9.6 THE SPACE-TIME TRAJECTORY OF THE PRODUCER SERVICE SUBGROUPS

In this section some spatial-dynamic aspects of the producer service sub-sectors will be considered. As stated in the previous section, there is a serious lack of data in the Netherlands on (comparable) time-series and spatial data with regard to the producer service subclasses. For example, statistics from which one could derive the space-time trajectory of the computer service sector are lacking.

Such an analysis would only, to some extent,[20] be possible at the 2-digit level (i.e, for SIC-84 as a whole). In this respect, The Chambers of Commerce are also unable to provide relevant data for each year concerning the spatial dispersion of the producer service subclasses.

Consequently, we will adjust our own sample in order to derive some, tentative conclusions. For this purpose the variable 'date of establishment', when the first activities in a certain location were started, will be crucial in our analysis. This variable will be used to approximate the spatial evolution of the subclasses distinguished. In this context we constructed the following *indicators* (which will be calculated for each of the three subgroups separately):

1. Regional share of employment (i.e., in December 1987) in establishments that have a 'date of establishment' earlier than January 1, 1976

2. The same with respect to a 'date of establishment' earlier than January 1, 1981

3. The same with respect to a 'date of establishment' earlier than January 1, 1985

4. Regional share of employment in establishments that have a 'date of establishment' later than December 31, 1984. This will also be denoted as regional share in 'new' employment.

Indicator 1 might, under certain conditions, be a reflection of the situation prevailing on December 31, 1975, and indicators 2 and 3 with respect to December 31, 1980 and December 31, 1984, respectively. To be sure that in a strict sense indicator 1, for example, is an adequate representation of the spatial dispersion of employment prevailing at December 1975, the following conditions have, inter alia, to be satisfied:[21]

a. Regional (total) growth of employment within establishments that already existed on December 31, 1975, and still existed in December 1987, should be the same in the period 1976-1987.

b. Regional 'death rates' for (total) employment within establishments, that existed on December 31, 1975 but stopped business later on, should be the same over this period.

c. Inter-regional relocation effects should cancel out during this period.

d. Regional 'shifting' of sectoral employment patterns due to establishments changing their main activity - i.e., changing their SIC-codes - should be equal.

It is rather difficult to determine the validity of these conditions, but it is clear that these conditions will become more problematic the longer the time interval considered and the smaller the regions analyzed. Some remarks however, seem to be appropriate in this respect. Firstly, since we have amalgamated the 40 COROP regions into 5 broad groups, these conditions are less restrictive than in the case of individual COROP regions.

Secondly, it appears that conditions b and c are especially biased against the approximated 'performance' of metropolitan areas on these dates. Concerning condition b, for example, Wever (1984) has demonstrated that the 'death rates' of new establishments appear to be the highest in metropolitan areas. Regarding condition c, the incubation hypothesis in its complex version (cf., Leone and Struyck 1976), as well as empirical research in the Netherlands (Buunk et al 1982, Molle et al 1982) has demonstrated that in the metropolitan areas a net out-migration of employment and establishments can be identified. On the other hand, the total effect of these relocations on regional employment patterns appears to rather small, however (cf., Bearse 1978, Potters 1983).

Thus, these studies might suggest an underestimation of the 'performance' of metropolitan areas on the basis of our above-mentioned indicators (on the prevailing dates). Concerning the 'direction' of the (possible) regional 'bias' of the conditions a and d no a priori judgements can be made, however.

Indicator 4 is of course intended to reflect the more *recent* spatial development patterns of our selected producer service subgroups. By means of this indicator it can be analyzed whether the various types of regions distinguished above are improving or worsening their relative 'performance' in the past three years.

The results of our analysis largely indicate the same type of space-time trajectory concerning the three producer service subclasses selected. Consequently, in this section we will discuss the patterns for the computer service sector in detail, while only some general remarks will be made concerning the 'general information' and 'engineering' subgroups (tables relating to these subclasses can be found in the appendix to this chapter). In the tables presented below the regional shares in total commercial service employment has also been recorded. The 'performance' of the various areas - concerning their shares of employment in the selected producer service subclasses - will be compared to these 'overall' shares, denoted here as 'normal' shares.

The results for the computer service sector can be found in Table 9.7 below.

Table 9.7 Approximated Space-Time Trajectory of the Computer Service
Sector

region	Approximated share of employment in pre-1976 establ. (indic. 1)	idem with respect to pre-1981 establishm. (indic. 2)	idem with respect to pre-1985 establishm. (indic. 3)	idem with respect to 'new' establishm. (indic. 4)	Approximated share of total employment Dec. 1987	Regional share commercial service employment SIC 6,7,8*
Metropolitan	64.5	62.3	56.2	50.8	54.9	44.8
'Overspill'	16.3	18.4	23.0	23.5	23.1	22.3
Brabant	9.2	11.0	11.6	13.2	11.7	12.2
Southern Peripheral	1.5	1.9	2.5	5.1	3.1	7.7
Northern Peripheral	6.1	5.2	5.5	7.0	5.8	13.0

Source: * Statistics of Employed
Persons 1986
Approximations do not sum to
100 because of rounding errors
and establ. with 'unknown'
location

If we accept indicators 1 - 4 as (rough) approximations of the spatial dispersion of employment prevailing on December 1975, December 1980, and December 1984, respectively, some interesting patterns can be found. In the first place, in the past the computer service sector appears to have been even more highly concentrated in metropolitan areas (a result also found in Koerhuis and Cnossen, 1982). As time passes by however, this dominant position is gradually being eroded, as suggested by the indicators used in Table 9.7. However, the share of 'new' employment - i.e., in establishments that have been set up in the period January 1, 1985 till December 1987 (and still existed in December 1987- still exceeds the 'normal' level[22] in the metropolitan areas. However, in this case the score on indicator 4 appears to be considerably below the scores on the other, historically oriented, indicators. These patterns of initial strong concentration in metropolitan areas versus (slow) spatial deconcentration also appear to be quite uniform among the individual 'metropolitan' COROP regions as reflected in Table 9.8 below.[23]

Table 9.8 Computer Service in Metropolitan Areas

region	Approximated share of employment in pre-1976 establ. (indic. 1)	idem with respect to pre-1981 establishm. (indic. 2)	idem with respect to pre-1985 establishm. (indic. 3)	idem with respect to 'new' establism. (indic. 4)	Regional share commercial service employment SIC 6,7,8*
Arnh./Nijmegen	8.4	5.7	5.1	2.8	4.1
Utrecht	19.5	17.8	18.5	16.0	7.9
Amsterdam	20.3	20.0	16.7	16.3	14.0
The Hague	11.4	8.1	7.0	6.7	6.8
Rijnmond	4.9	10.7	8.9	9.0	12.1

Source: * Statistics of Employed Persons 1986

Our indicators suggest that the Rijnmond Area performs relatively poorly compared to the other metropolitan regions distinguished. In the Rijnmond case, none of our indicators exceeds the 'normal' level, taken to be its share in total commercial service employment. Also concerning its share in 'new' employment - i.e., indicator 4 - the Rijnmond area still performs rather poorly.

As far as the share in 'new' employment is concerned, the Utrecht area still appears to attract a share that exceeds its 'normal' level considerably.[24] Also the Greater Amsterdam Area still succeeds in attracting a higher than 'normal' share of 'new' employment. The results in Table 9.8 above also indicate the important (leading) role the Greater Amsterdam Area has played in the 'take-off' of the new (and innovative) computer service sector in the Netherlands (see also Chapter 5).

The second type of region distinguished in our analysis, which seems to have passed its 'normal' level, appears to be the *'overspill'* *area* which by 1985 (as reflected by indicator 3) has apparently passed its 'normal' share of computer service employment.

The third region in order appears to be the intermediate *Brabant region*. Although until December 1987 this area did not yet attract a 'normal' share of computer service employment, this region is gradually improving its 'performance'. In this context, this area has also attracted an 'above-normal' fraction of 'new' employment (indicator 4).

On the other hand, the *southern* and (especially) the *northern peripheral areas* have not yet succeeded in attracting a 'normal' share in computer service employment. Their shares in 'new' employment (indicator 4) are also still considerably below their shares in total commercial service employment. In comparing the performance of these areas on the indicators 1 till 4 however, the southern peripheral area appears to be more rapidly expanding its share than the northern peripheral area.

Consequently, the space-time trajectory of the computer service sector, as approximated by our indicators, appears to be clearly developing along the lines of our theoretical framework for sectors linked to new 'technology systems' sketched in the first part of this book. In comparing the indicators 1 to 4 in Table 9.7 above it is clear that at first a very strong metropolitan concentration can be observed, while in later phases the pattern seems to be gradually shifting towards less central (metropolitan) areas.

Next, we will briefly discuss the approximated time-space trajectory for the *'engineering'* subgroup (see Tables 9.9 and 9.10 in the appendix). This subgroup appeared to be the most 'static' sector of our selected subgroups, i.e., in terms of the generation of 'new' establishments (cf., Table 9.5). Consequently, we expect the space-time trajectory of this sector - along the lines of our 'dynamic incubation' framework - to be further developed than for the other sectors selected.

This is confirmed by the data as regional shares appear to be less fluctuating (compared to computer service and 'general information': see below), while these shares are also more closely related to the 'normal' shares (see columns 6 and 7 of Table 9.9).

In the course of time, the 'metropolitan' areas (again) appear to be continuously losing their dominant position, while especially the 'overspill' and Brabant areas appear to be improving their relative 'performance'. As can be concluded from Table 9.10 in the appendix this deconcentration pattern also applies quite uniformly to the 'metropolitan' COROP regions separately.

In this context, all individual 'metropolitan' COROP regions distinguished attract a smaller proportion of 'new' employment (indicator 4) compared to their share of pre-1985 employment (indicator 3).[25] As regards the regional shares in 'new' employment the Amsterdam, Rijnmond and Arnhem/Nijmegen areas are nowadays already performing below their 'normal' levels (see Table 9.10).

While the *southern peripheral area* appears to be gradually improving its performance, developments in the *Brabant* and '*overspill*' areas seem to be more explosive. Concerning their share of 'new' employment, both the Brabant and 'overspill' areas attract a proportion which is larger than their shares in total commercial service employment. The northern peripheral region scores reasonably well on the 'historical' indicators (i.e., 1 to 3). However, the score of this region on the 'new' employment indicator 4 is below 'normal'.

Consequently, our indicators suggest that also concerning the engineering 'subgroup' the metropolitan areas as a whole experience a gradual erosion of their dominant position. The opposite pattern holds for the 'overspill' and Brabant region. The southern peripheral region is also gradually improving its performance. The northern peripheral area has a 'normal' score on the 'historical' indicators 1,2, and 3. However, the score of this region on indicator 4 indicates a deterioration of this rather favourable position.

Lastly, we will consider the approximated space-time trajectory for the 'general information' subgroup (see Tables 9.11 and 9.12 in the appendix).

As found for the other subgroup, the 'general information' subgroup appears to be heavily concentrated in metropolitan areas as well. Again however, in the course of time this dominant position seems to decline (see indicators 1-4 in Table 9.11).

Looking more specifically at the individual metropolitan areas, Table 9.12 in the appendix shows that especially the Greater Amsterdam region appears to have been an important 'incubation area' for this sector (see the historical indicators 1-3 in Table 9.12). However, the results for indicator 4 point at a deterioration of this dominant position. As far as 'new' employment is concerned, this region only attracts a 'normal' fraction as reflected by indicator 4 in Table 9.12.

The other metropolitan areas on the other hand - with the exception of the Greater Rijnmond area - still appear to be in a kind of 'boom' phase. As can be seen from Table 9.12 in the appendix, the shares of the COROP regions Arnhem/Nijmegen, The Hague and Utrecht in 'new' employment are considerably larger than their shares in pre-1985 employment.

Also concerning the 'general information' subgroup the 'overspill' and Brabant areas hold an 'intermediate' position. The Brabant region already appears to have attracted a 'normal' share of employment in this subsector, while also the 'overspill' area has nearly attained its 'normal' level. As far as 'new' employment is concerned, this region appears to have attracted a larger share than its share in total commercial service employment.

The peripheral areas on the other hand, although also expanding, again exhibit considerable below-normal scores on our indicators. Even concerning the shares in 'new' employment (indicator 4) both the northern and southern areas still perform relatively poorly.[26]

Consequently, the spatial patterns found for the 'general information' subgroup appear to be largely consistent with the patterns found for the computer service and engineering subgroup. In the next section, the most important conclusions of this chapter will be recapitulated.

9.7 CONCLUSION

In this chapter our main purpose has been to trace the spatial dispersion of relatively new (and innovative) sectors linked to the more recent 'information technology system'. In this respect, we have in particular focussed attention on the producer service sector. This sector is generally acknowledged to be a highly dynamic (and innovative) sector nowadays.

As to the empirical testing of our theoretical framework developed in the first part of this book, we have tried to determine whether the spatial development pattern of selected subclasses of this sector displays a hierarchical spatial diffusion pattern as suggested by the second hypothesis deduced from this framework.

To this purpose, first of all the increased economic importance of the producer service sector as a whole - e.g., in terms of value-added and employment - for the Dutch economy has been demonstrated. Next, producer service subgroups have been selected which can be considered to be most closely related to the more recent 'information technology system'.

For an identification of the spatial dispersion of (employment in) these subgroups, the lack of relevant data forced us to construct our own data set by means of information provided by the Chambers of Commerce. From these data it appeared that especially the computer service, 'general information' (e.g., management consultancy) and the engineering subgroup can indeed be considered as highly dynamic - i.e., in the sense of many new ('Schumpeterian') entrances - components of the producer service sector.

Consequently, as to the empirical testing of our theoretical framework - and in particular the second hypothesis deduced from this framework - we have analyzed the spatial dispersion of these three subgroups. In this context, it has been demonstrated that until recently employment in each of the highly dynamic subgroups distinguished appears to be heavily concentrated in metropolitan areas.

In this respect, the (semi-)intermediate areas - i.e., the 'overspill' and Brabant areas - in general hold an 'intermediate' position. The peripheral areas on the other hand, appeared to perform relatively poorly (the only exception being the 'score' of the northern peripheral zone with respect to the engineering subgroup). Consequently, these spatial patterns are largely consistent with the fourth hypothesis mentioned above.

However, in order to further test the fourth hypothesis, besides the *static* spatial patterns we have also attempted to approximate the *time-space* trajectory of the selected subgroups. In this respect, the main purpose has been to determine whether in the course of time spatial deconcentration tendencies can be observed. To this purpose, given the lack of relevant data, several proxy indicators have been constructed which might serve as a reflection of the spatial dispersion of employment in these subgroups at certain moments in time. In this manner we have tried to approximate the time-space trajectory of the selected producer service subclasses.

The empirical results largely support the fourth hypothesis mentioned above concerning the spatial diffusion of new 'Schumpeterian' sectors. In former periods, each of the subgroups distinguished appears to have been even more heavily concentrated in metropolitan areas. In later phases however, this dominant position seems to diminish gradually while the other (non-metropolitan) regions are improving their relative performance.

As regards these regions the (semi-)intermediate areas in general appear to be first in attracting a 'normal' share of employment in the subclasses distinguished. The peripheral areas on the other hand, clearly lag behind in adopting these sectors and in succeeding to attain a 'normal' fraction of employment in these subgroups (the only exception being the performance of the northern peripheral areas with respect to the engineering subgroup).

As far as the performance of the Greater Amsterdam Area is concerned (see Chapter 5), it has been demonstrated that this region has played an important (leading) role in the 'take-off' of both the computer service and 'general information' subgroup.

Consequently, the (approximated) time-space trajectories of our selected subgroups largely display a centre-periphery pattern as suggested by the fourth hypothesis deduced from our theoretical framework. This spatial diffusion pattern appears to be the farthest progressed for the most 'static' (in terms of new establishments) of these subgroups, i.e., the engineering subgroup.

Concerning the most 'dynamic' sub-sector analyzed in this chapter - i.e., the computer service subgroup - employment still appears to be most heavily concentrated in metropolitan areas. The peripheral areas on the other hand, perform most poorly concerning this subgroup.

This chapter concludes our empirical investigations. In the next chapter, the most important findings of this book will be recapitulated.

Notes to Chapter 9:

1. In these statistics self-employed persons are excluded.

2. This follows from the increased importance of part-time jobs.

3. This is caused by the fact that statistical revisions had a rather severe impact on the producer service statistics.

4. Fortunately SIC-85 is only small in terms of employment (i.e., about 2 %) compared to SIC 84.

5. The growth of the producer service sector as a whole is often (partly) ascribed to 'crowding out' or an 'externalization effect'. This implies that firms which previously performed such services in-house now board out these services, because of either cost or expertise motives. In the first case, costs would be higher should firms perform these services in-house. In the second case however, the lack of expertise with respect to the tasks needed is the main driving force for externalization.

6. This applies especially to research performed at the level of individual establishments, as was also confirmed in a comparative study by SKIM (1981).

7. This does not imply however that this data base can be considered to be error-free (cf., Wever 1984). During 1984 however, the Central Databank considerably improved the quality of its data base by removing non-existent establishments, by improving SIC-coding.

8. In this respect, both main and subsidiary establishments have been included in the sample.

9. Considering these (May, 1987) figures as the total population of a repeated Bernoulli experiment, the sample outcomes - representing the December 1987 situation - indicate that also for these size categories the total number of establishments appears to have increased considerably in the period May-December 1987. This can be understood by reconsidering the first two moments of the Binomial distribution (our sample design for these small establishments can be considered as a repeated Bernoulli experiment). If N=Total population size and if we have a repeated Bernoulli experiment in which every member of the population has a chance p of being selected (and 1-p of not being selected) we have:
$E(x)$= expected outcome (i.e., number of successes) = $p_* n$
$V(x)$= variance of outcome   = $p_* (1-p)_* n$

Taking computer-service as an example and considering the establishments with one employee, we have:

N=2352 (i.e., the total May figure of the Databank)
p=0.1 , E(x)=235, V(x)=211

By constructing a 95% confidentiality interval around E(x) (i.e., 235 $\pm$ 1.96*14.5), it can be seen that an outcome of 263 (as we observed in our sample) would be highly implausible given (a supposed) total population of 2,352 (i.e, the May 1987 situation). This holds even more so for the observed sample sizes within the general information sector (for both size categories) as well as the computer-service sector (for establishments with 2-4 employees). Consequently, our conclusion must be that especially with respect to these subsectors - concerning all three size categories distinguished - the total number of establishments has increased during this relatively short period.

10. Given the sample fractions for the smaller size categories (see the second column of Table 9.4) the number of establishments in the size categories 1 and 2-4 employees - as observed in our sample - have been multiplied by 10 and 3, respectively.

11. A small fraction of this may be ascribed to establishments changing their location to a different Chamber of Commerce which results in a new date of establishment. In general this effect will only be minor (cf., Wever 1984), especially as we are only considering a relatively short time period (less than 3 years).

12. This might be considered as a proxy for the degree of success of 'new' establishments, i.e., when they succeed in passing the 'critical' threshold of one employee (i.e., self-employed persons).

13. For the subsector 'architect offices' similar estimations could be constructed which indicated a pattern even 'worse' - as far as the percentages of 'new' establishments is concerned - than for the financial information subgroup. As also these offices are included in SIC 84, this is again a clear illustration of the rather heterogeneous nature of the sub-sectors included in this 2-digit SIC category.

14. Although belonging to the intermediate zone of the Netherlands (as discussed before), also the Arnhem/Nijmegen region has been included in this 'metropolitan' subgroup. As stated in Chapter 5, this region is largely metropolitan while the development of manufacturing employment in this region largely corresponds to the other 'metropolitan' COROP regions (located in the central zone of the Netherlands).

15. Although not really an 'overspill' area of the Western Netherlands, this COROP region has been included in the second group because the observed patterns appeared to be rather consistent with this group and inclusion in the other sets may be even more unrealistic. In any case, this choice is rather irrelevant for the empirical results that will follow.

16. In making these estimations we multiplied the observed employment figures from the sample with a factor 10 and 3 respectively for establishments with 1 and 2-4 employees (being equal to the inverse of the chances of being selected: see column 2 of Table 9.4).

17. As none of the establishments in the selected subclasses exceeded the 'critical' threshold of 1,000 employees, the size category of 1,000 and larger posed no problem in this respect.

18. These figures have been derived from the Statistics of Employed Persons 1986. These statistics exclude self-employed persons.

19. The results presented in Table 9.5 indicate that this sector is also the most 'static' from our selected subgroups.

20. By using the Statistics of Employed Persons which does not count the number of self-employed persons, however.

21. Similar conditions should hold for indicators 2 and 3, only for shorter periods of time (i.e., for the periods December 1980-December 1987 and December 1984-December 1987, respectively).

22. As stated before, the term 'normal' level refers to the regional shares in total commercial service employment.

23. As said before, one should be more careful in interpreting the data at such a disaggregate spatial level as especially the historically oriented indicators may become less reliable in this respect.

24. Although not exactly comparable in terms of relevant periods, the results of Koerhuis and Cnossen (1982) concerning 1981 suggest that our approximations concerning the 'performance' of the metropolitan areas as a whole in December 1980 (indicator 2), as well as the individual results for the Amsterdam and Rijnmond area, are very reasonable indeed. On the other hand, the performance of the individual Utrecht area with respect to indicator 2 may have been overestimated, while that of the individual The Hague region may have been underestimated. This would imply that we are under-estimating the 'improved performance' of the former region, while under-estimating the relative decline of the latter region in the recent decade.

25. In case we would exclude COROP region Delft en Westland from our analysis with respect to indicator 4 (for an explanation: see the appendix), this conclusion would not hold for the COROP regions Utrecht and The Hague.

26. In this context, the score of the northern peripheral area on indicator 4 can especially be ascribed to the province of Groningen in which the city of Groningen is located (4.9% of total 'new' employment in 'general information' is estimated to be generated in this province).

Appendix 1

Table 9.9 Space-Time Trajectory of the Engineering Sector*

Region	Approximated share of employment in pre-1976 establ. (indic. 1)	idem with respect to pre-1981 establishm. (indic. 2)	idem with respect to pre-1985 establishm. (indic. 3)	idem with respect to 'new' establishm. (indic. 4)	Approximated share of total employment Dec. 1987	Regional share commercial service employment SIC 6,7,8*
Metropolitan	60.1	55.9	51.9	42.2 (48.6)	51.1	44.8
'Overspill'	16.2	17.5	19.0	30.6 (20.1)	20.0	22.3
Brabant	5.1	7.2	9.3	12.6 (14.5)	9.9	12.2
Southern Peripheral	4.4	4.9	4.9	5.6 (6.4)	5.0	7.7
Northern Peripheral	14.2	14.5	14.5	8.8 (10.0)	14.0	13.0

Source: * Statistics of Employed
Persons 1986
Approximations do not sum to
100 because of rounding errors
and establ. with 'unknown'
location

Table 9.10 Engineering in Metropolitan Areas*

Region	Approximated share of employment in pre-1976 establ. (indic. 1)	idem with respect to pre-1981 establishm. (indic. 2)	idem with respect to pre-1985 establishm. (indic. 3)	idem with respect to 'new' establishm. (indic. 4)	Regional share commercial service employment SIC 6,7,8*
Arnh./Nijmegen	9.6	7.9	6.6	3.1 (3.6)	4.1
Utrecht	9.0	8.8	9.7	9.1 (10.5)	7.9
Amsterdam	14.3	11.8	10.0	8.6 (9.9)	14.0
The Hague	14.9	14.0	12.5	11.7 (13.5)	6.8
Rijnmond	12.3	13.4	13.0	9.7 (11.2)	12.1

Source: * Statistics of Employed
Persons 1986

As the COROP region Delft attracted a few large engineering establishments in recent years, the 'overspill' region scores relatively favourable on indicator 4. Consequently, in order to exclude this potential 'bias', this indicator has also been re-estimated by excluding this region. These estimations (which imply an 'over-estimation' of the performance of the other regions and an 'under-estimation' of the 'overspill' region) have been presented in parentheses in column 5 of Tables 9.9 and 9.10

Table 9.11   Space-Time Trajectory of 'General Information'

region	Approximated share of employment in pre-1976 establ. (indic. 1)	idem with respect to pre-1981 establishm. (indic. 2)	idem with respect to pre-1985 establishm. (indic. 3)	idem with respect to 'new' establishm. (indic. 4)	Approximated share of total employment Dec. 1987	Regional share commercial service employment SIC 6,7,8*
Metropolitan	64.8	60.1	56.5	48.7	54.6	44.8
'Overspill'	16.8	18.0	19.6	23.4	20.6	22.3
Brabant	10.6	12.3	13.0	14.4	13.3	12.2
Southern Peripheral	2.5	3.1	3.5	3.1	3.5	7.7
Northern Peripheral	5.1	6.5	7.1	9.1	7.7	13.0

Source: * <u>Statistics of Employed Persons</u> 1986
Approximations do not sum to 100 because of rounding errors and establ. with 'unknown' location

Table 9.12   'General Information' in Metropolitan Areas

region	Approximated share of employment in pre-1976 establ. (indic. 1)	idem with respect to pre-1981 establishm. (indic. 2)	idem with respect to pre-1985 establishm. (indic. 3)	idem with respect to 'new' establishm. (indic. 4)	Regional share commercial service employment SIC 6,7,8*
Arnh./Nijmegen	1.1	2.2	2.5	5.0	4.1
Utrecht	7.9	8.8	9.4	12.0	7.9
Amsterdam	38.1	32.1	28.3	13.7	14.0
The Hague	5.2	6.2	6.2	8.5	6.8
Rijnmond	12.5	10.8	10.1	9.5	12.1

Source: * <u>Statistics of Employed Persons</u> 1986

10 Conclusion

10.1 RETROSPECT

In this study on incubation and innovation new economic activities have been analyzed from a spatial perspective. Based on an extensive literature survey in the first part of the book, two existing basic research 'gaps' have been further analyzed in the present study. The first basic research issue identified concerns the impact of locational variables - or the (latent) variable 'production milieu'- upon the 'innovativeness'[1] of individual firms or establishments. Thus the question was here: does the innovativeness of firms - in addition to their intra-firm characteristics - depend on their location in qualitatively different production environments?

Various innovation studies have indicated regional variations in the innovativeness of individual firms. However, empirical evidence based on firm analytical grounds considering the impact of both intra-firm and production milieu characteristics simultaneously upon regional variations in the generation of innovations is very scarce. Given these considerations, we have tried to determine in the second (empirical) part of this study the impact of both intra-firm and production environment characteristics upon the innovativeness of various types of Dutch manufacturing establishments.

A second research question brought to the fore by our literature survey concerned the general lack of a theoretical conceptual framework concerning the spatial evolution pattern of (new) life cycles of industries and technologies. In this context, it has been observed that *empirical* spatial innovation research - relating to changes in technology and regional structure - has progressed more firmly than *theoretical* analysis.

Consequently, in the first part of this study an attempt at designing such a theoretical - but nevertheless operational - framework has been undertaken. In this so-called dynamic incubation framework, more recent approaches regarding the role and characteristics of technological change in economic analysis have been incorporated with their spatial dimension. In this context, we have particularly called attention to the role of permanent technological change in shaping and rearranging spatial structures. Based on a macroscopic and long term view, our attention has mainly focused on the spatio-temporal dynamics related to the emergence of new 'technology systems' and the resulting new life cycles of industries and technologies.

In this framework technological change - relating to the new types of products and services opened up by the emergence of a new 'technology system' - has been conceived of as a 'creative diffusion' process. In this respect, the 'swarming processes' of (new) 'Schumpeterian' firms - attracted to the high (initial) profit potential of these new types of products and services - frequently involves the generation of further product and process innovations resulting in new life cycles of industries and technologies. The 'paths' of further innovative activities regarding such new types of products and services have been denoted as technological 'trajectories'. As far as the types of (further) innovations generated along newly established technological 'trajectories' are concerned, the main emphasis in the course of time will likely shift from product to process oriented innovations (denoted here as the innovation life cycle concept).

As far as the *spatial* dimension is concerned, four (testable) hypotheses may be deduced from the above framework. Firstly, metropolitan or central areas are expected to be leading in the structural economic transformation processes resulting from the emergence of new 'technology systems'. Or - stated otherwise - such regions are expected to take the lead in both the *growth* and the *decline* of life cycles of industries and technologies. Secondly, this framework suggests a 'hierarchical' spatial diffusion pattern of new

('Schumpeterian') sectors related to the emergence of new 'technology systems'. In general several types of regions are expected to play their own specific role in the 'creative diffusion' process mentioned above. In this respect, a third hypothesis suggests that in later phases of the life cycles of industries and technologies non-metropolitan or non-central areas may become dominant concerning the innovative activities performed. And finally - given the increased importance of the generation of process innovations in these later phases - this 'dominance' will especially apply to the generation of process innovations. Such interesting hypotheses require of course a further empirical test. This issue has been considered in the second part of this study.

10.2 RESULTS ACHIEVED

In the empirical part of this study an attempt has been made to examine the empirical validity of the theoretical framework - developed in the context of the second basic research issue identified above - by means of testing the four hypotheses mentioned above. In this part, the first basic research issue on the impact of the production milieu upon entrepreneurial innovativeness has also been considered.

This research issue was analyzed for various types of Dutch manufacturing establishments - labelled firms here - by using the results of an extensive industrial survey. In order to analyze specific selection environment impacts, various regional indicators - assumed to reflect 'high quality' production environments - were considered. By using a latent variable approach it was demonstrated that spatial variations in the innovativeness of manufacturing firms appear to be especially correlated with differences among these firms in terms of their innovation potential measured by means of intra-firm characteristics. After correcting for this 'structural' (intra-firm) component, an additional positive 'production milieu' impact upon the innovativeness of these firms - resulting from a location in a 'high quality' production environment - could, however, not be gathered from the dataset.

In the second part of this study the empirical validity of our theoretical framework was further analyzed by means of testing the four hypotheses mentioned above.

In order to test the first hypothesis on the leading role of large metropolitan areas, as a case study we analyzed the structural economic changes in the Amsterdam area. The patterns observed for this area correspond largely to the first hypothesis mentioned above. As far as the decline of employment in the manufacturing sector is concerned the Amsterdam area clearly preceded the decline for the Netherlands as a whole (see also below).

It is noteworthy here that the relative poor performance of this area regarding the manufacturing employment in recent decades, appeared to be also reflected in a weak innovation potential of firms (which confirms the third hypothesis; see also below). On the other hand however, it was also demonstrated that the Amsterdam area has played an important leading role in the 'take-off' of various highly dynamic producer service (sub)sectors (see also below). The Amsterdam area appears to be a forerunner in a structural transformation process towards a post-industrial society. A necessary condition for maintaining this position is of course the removal of bottlenecks (e.g., infrastructure) which might hamper a smooth adjustment and flexible response to new challenges.

As regards the empirical validation of the second hypothesis, the spatial dispersion of relatively new (and innovative) producer service (sub)sectors - linked to the more recent information 'technology system' - was also analyzed for the Dutch context. From this analysis it appeared that until recently these new dynamic activities were indeed heavily concentrated in metropolitan areas. On the other hand, these areas also appeared to be leading as regards the decline of (employment in the) manufacturing sector (which confirms the first hypothesis; see also below).

However, our analysis also provided several indications that the spatial concentration mentioned above has been especially important in earlier phases, while in later phases spatial deconcentration appeared to come to the fore. In this respect, the (semi-)intermediate areas in general appear to be followers in succeeding to attract a 'normal' share of employment in these relatively new sectors, while peripheral areas take up a rear position. Consequently, these hierarchical spatial 'diffusion' patterns largely confirm the second hypothesis mentioned above.

As stated above, the manufacturing sector rather rapidly lost its employment generating capacity in recent decades. Generally speaking the lack of new technological 'trajectories' together with a decline in the possibilities to generate (further) product innovations along previously established 'trajectories' had a negative impact upon employment in these periods. In this context, the generation of (labour-saving) cost-efficient process innovations became increasingly a critical factor.

In this respect, especially the central (metropolitan) regions appear to have suffered from this decline in manufacturing employment. In our theoretical framework - see the third hypothesis - it has been hypothesized that in such later life cycles of industries and technologies the central (metropolitan) areas will also lag behind as regards the innovative activities performed.

In order to test the empirical validity of this hypothesis we analyzed the performance of various (types of) Dutch regions - i.e., on the basis of the industrial survey mentioned above - regarding the innovation potential of its firms. This analysis has been performed for several types of (small) manufacturing firms. The results of our analysis indicate that the innovation potential of several (metropolitan) regions in the central zone of the Netherlands is relatively weak for various types of firms analyzed.

Finally, as far as the types of innovations are concerned - see the fourth hypothesis - our empirical results suggest that the central zone of the Netherlands especially lags behind in the generation of (high quality) process innovations, but not in the generation of (high quality) product innovations.

Consequently, the empirical findings are encouraging as regards the empirical validity of the theoretical framework developed in the first part of this study. Clearly, further theoretical/methodological refinements and additional empirical analyses are desirable in order to understand the complex interrelations between technological change and spatial dynamics; a number of which will be discussed subsequently.

10.3 POLICY IMPLICATIONS

Our analysis contains various aspects which may be relevant from a policy viewpoint. As stated above, our empirical results suggest that the impact of location in high quality production environments upon the innovativeness of individual (manufacturing) firms is - ceteris paribus the intra-firm characteristics - rather limited. Although undoubtedly further research efforts are desirable in this context (see also below), these results imply a warning against overly optimistic expectations of regional policies aiming at increasing the innovativeness of local firms by means of 'upgrading' the local production environment. At least in the short run, the impacts of such policies may prove to be rather limited.

This does of course not imply that such policies are irrelevant from a long-term perspective as in the long-run both the 'production milieu' and intra-firm characteristics may be interrelated. However, the long-run impact of changes in the attributes of the local production environment upon the internal - i.e., innovation potential- characteristics of (local) firms is still a field of scientific and policy interest open to further research.

As to the empirical determination of this impact, especially a longitudinal analysis appears to be a promising analyis framework (see, for instance, Van der Knaap et al 1988). By means of such an analysis further insights into the linkages between various elements of the local production environment and internal characteristics of firms could be provided. Such insights are also of utmost importance to an efficient allocation of policy means aiming at improving the indigeneous innovation potential.

The empirical results provided in this study indicate that especially the 'structural' or intra-firm component is important in 'explaining' regional variations in innovativeness of manufacturing firms. Or - stated otherwise - regional differences in the firm-specific innovation potential appear to be important in this context. Consequently, these results seem to indicate that regional policies aiming at improving the innovativeness of local firms should especially concentrate on this 'structural' component. Clearly, this requires a long-term planning horizon focusing on strategic issues.

In this respect, two broad alternatives come to the fore. Firstly, regional policies aiming at attracting or generating new 'high tech' firms may be considered as one meaningful alternative. As discussed in this study however, this alternative may only be open to a rather limited set of metropolitan regions as such 'firms' are often assumed to be highly dependent upon facilities (e.g., a high skilled labour force, educational and training facilities, communication networks) provided by such regions.

Regional policies aiming at increasing the indigenous innovation potential of *existing* firms appear to be a second alternative in this respect. As regards this second alternative, the 'upgrading' of the local production environment is a first possibility. As stated above however, still little empirical knowledge exists about the (long-run) impact of such policies upon the intra-firm innovation potential.

As a second possibility concerning this alternative, regional policy efforts may try to influence the internal - i.e., innovative-characteristics of local firms in a more direct way (e.g., by increasing their R&D efforts). This would require a tailor-made policy, in terms of both regional orientation and firm orientation. However, such specific policy measures are (still) unavailable in many countries; most traditional regional or innovation policies are focusing attention on more generic instruments (e.g., new technology subsidies; see also Davelaar and Nijkamp 1987e).

However, a meaningful direction may be found by removing barriers in the transfer of knowledge regarding new technologies and their potential applications. Policies favouring regional-oriented science-parks or regional innovation centres - favouring custom-made knowledge transfer to firms - may be helpful (see for instance Premus 1982, Malecki and Nijkamp 1988), provided at least that such parks or centres do not become mutual competitors fighting for a maximum share of a given market.

To avoid such a zero-sum game strategy, an exploitation of a distinct regional development potential would be necessary. Such a custom-oriented policy is developed at present in the Netherlands and may contribute to an improvement of the indigenous innovation potential, provided the policy initiatives are geared toward the carrying capacity of the regions concerned (e.g., in terms of social overhead capital, skilled labour force; see Nijkamp 1988).

A related policy measure might be to stimulate the indigenous innovation potential of regions via government purchase policy or research contracts. This is comparable to the US government policy by means of research contracts for defence. In such case the risks inherent in the generation of innovations are reduced for the regions concerned.

Our framework sketched above also has implications for the well-known 'efficiency-equity' dilemma in regional policy. Because of the embodied 'creative diffusion' process, our framework reflects a more optimistic point of view for non-metropolitan areas than is usually suggested by more conventional theoretical approaches in spatial analysis (e.g., the filtering-down theory, the spatial product life cycle approach).

Implicitly, these approaches assume that technological progress concerning new types of products or services has come 'at rest', before the production shifts - because of a-technological factors such as labour costs, pollution regulations, unionization - to non-metropolitan areas. This ignores the creative and innovative role - and the resultant competitive power - of these non-central areas in later life cycles of industries and technologies, however.

In our framework, the 'creative diffusion' process lies at the heart of the spatial shift of both - mutually reinforcing processes of - production *and* innovation in these later phases (and not only of production as perceived in the above mentioned approaches). Also our empirical results demonstrate quite clearly that - as far as the Netherlands is concerned - the generation of innovations is not restricted to metropolitan or central areas.

In our framework several types of regions are expected to play their own specific innovative role in a 'creative diffusion' process concerning new types of products and services opened up by the emergence of new 'technology systems'. This implies that the long-run sustainability of these different types of regions is interrelated and complementary rather than competitive. On the one hand, metropolitan regions provide the seedbed function for the 'take-off' of new

('Schumpeterian') sectors related to more recent 'technology systems'. On the other hand however, in later phases the inherent dynamics will shift to other (non-metropolitan) areas.

In this context, an explicit and active policy intervention in favour of 'equity' - i.e., in the sense of an equal 'industry mix' in each region - might even disturb these symbiotic relations. This policy aim might even prove to be incompatible with the long-run sustainability of these different types of regions. On the one hand, metropolitan areas - because of the high bid-rents prevailing here- are largely dependent upon sectors related to more recent 'technology systems' which are generally able to generate a high value-added. On the other hand, non-metropolitan areas are generally unable to provide the necessary seedbed conditions important in earlier life cycles of industries and technologies (cf., Malecki and Nijkamp 1988).

In the context of our analysis, a twofold approach toward an 'optimal' policy strategy for regional innovation is suggested. Firstly, to warrant and utilize the incubation function of large metropolitan areas concerning the 'take-off' of new economic activities. Secondly, to promote the adoptive potential of other areas - and further innovative activities to be performed here - in later phases. Of course, further regional innovation policy research would be needed in order to validate these assumptions.

10.4 NEW RESEARCH AREAS

In view of the above observations, various new directions for future research efforts may be envisaged. Five such core research areas will briefly be discussed here.

As a first field of scientific interest we already mentioned the lack of empirical evidence concerning the long-run impact of (various attributes of) the local production environment upon the internal- i.e., innovative - characteristics of firms.

In the context of our analysis, a second fruitful research alley consists of gaining further empirical insights into the interrelations between life cycles of relatively new ('Schumpeterian') sectors and regional growth cycles. In this respect, a longitudinal-sectoral approach may be meaningful. By analyzing different types of regions in such an analysis, one might try to determine the direct and indirect regional growth effects resulting from developments within various producer service subgroups. Such an analysis can also provide further insights into the development of interregional growth differentials. In this way the empirical validity of the theoretical framework developed in the first part of this study can be the subject of further empirical tests. This would especially be the case if the regional innovative component within these subgroups would be taken into consideration as well (see also the final research suggestion mentioned below).

Thirdly, the growing importance of information inputs in economic production processes is nowadays generally acknowledged. As pointed out, this has largely affected the growth pattern of various producer service (sub)sectors. However, also within several manufacturing sectors the substitution process of 'blue collar' for 'white collar' workers becomes increasingly important. At the spatial level the question can be raised whether (the speed of) these intra-sectoral substitution processes differ for various types of regions. Theoretical and empirical analysis concerning this question is still rather scarce, however. The empirical answers to this research issue are also highly relevant for (future) labour, education and communication requirements in the various regions concerned. Consequently, a completion of this research 'gap' appears to be of utmost importance from the viewpoint of the long-run regional development potential and regional policy efforts.

A fourth important research area may be found in the field of analyzing specific bottlenecks of firms with regard to their innovative activities, both in sectoral and regional terms. Such an inventory may also be important in view of the coordination of regional innovation policies (e.g., cooperation between the regional innovation centres). This may also be favourable from the viewpoint of region-specific development policies.

Finally, the analysis of innovative activities performed within various commercial service sectors deserves further attention. Although it is generally acknowledged nowadays that also within these sectors innovations are being generated, the mere measurement of the innovation potential and/or innovativeness of firms operating in such sectors is still underdeveloped. This also holds for the determination of the regional discrepancies in the innovative activities performed within these sectors. Consequently, as regards this research issue, there is a lack of appropriate 'measuring rods' to gauge the concepts mentioned above.

In general, the area of the spatial incubation and innovation dimensions of new technologies appears to be a rich research field. In view of the restructuring of the European economy this issue will no doubt gain further importance in the near future. The modest attempts made in the present study suggest at least that this rapidly evolving research area is likely to show a promising and fruitful life cycle.

Notes to Chapter 10:

1. As stated before, innovation potential refers to the production factors for innovative activities performed by firms (e.g., R&D efforts), while innovativeness refers to the output of these activities (e.g., number of product innovations generated).

References

Abernathy, W.J. and J.M. Utterback (1978), Patterns of Industrial Innovation, *Technology Review*, vol. 80

Alderman, N. and A. Thwaites (1987), Research and Development and Technical Change in Traditional Industries: A Regional Perspective, Paper presented at the 27th Regional Science Association European Congress, Athens, August 25-28, 1987

Alderman, N. (1985), Predicting Patterns of Diffusion of Process Innovations within Great Britain, paper presented to the twenty-fifth European Congress of the Regional Science Association, Budapest, Hungary, 27-30 August, 1985

Alderman, N., P. Wynarczyk and A.T. Thwaites (1986), High Technology, Small Firms and Regional Economic Development : A Question of Balance, Paper presented at the International Workshop on the Role of Small and Medium Size Enterprises in Regional Development, SAMOS, september 25-27, 1986

Alders, B.C.M. and P.A. de Ruijter (1984), De Ruimtelijke Spreiding van Kansrijke Economische Activiteiten in Nederland, TNO/PSC, Delft/Apeldoorn

Allen, P.M. and M. Sanglier (1979), A Dynamic Model of Growth in a Central Place System, *Geographical Analysis*, vol.11, pp. 256-272.

Andersson, A. and B. Johansson (1984), Knowledge Intensity and Product Cycles in Metropolitan Regions, Contributions to the Metropolitan Study, no.8, IIASA, Laxenburg

Apel, H. (1980), Soft Models of Herman Wold's Type: Models of Environmental Problems and Underdevelopment, in: W. Buhr and P. Friedrich (eds), *Regional Development under Stagnation*, pp. 259-279, Nomos Verlag, Baden-Baden

Areskoug, B. (1982), The First Canonical Correlation. Theoretical PLS Analysis and Simulation Experiments, in: K.G. Jöreskog and H. Wold (eds), *Systems under Indirect Observation, Causality, Structure, Prediction*, vol. 2, pp. 259-279, North-Holland, Amsterdam

Aydalot, P. (1984), Reversals of Spatial Trends in French Industry since 1974, in: J.G. Lambooy (ed), *New Spatial Dynamics and Economic Crisis*, pp. 41-62, Finnpublishers, Tampere

Ayres, R.U. (1987), Barriers and Breakthroughs: An 'Expanding Frontiers' Model of the Technology-Industry Life Cycle, IIASA, Laxenburg.

Bade, F.J. (1986), The Deindustrialization of the Federal Republic of Germany and its Spatial Implications, in P. Nijkamp (ed), *Technological Change, Employment and Spatial Dynamics*, pp. 196-220, Springer Verlag, Berlin

Baker, R.J. and J.A. Nelder (1978), *The GLIM System. Release 3. Generalised Linear Interactive Modelling*, Rothamsted Experimental Station, Harpenden, Herts, England

Barkley, D.L. (1988), The Decentralization of High-Technology Manufacturing to Nonmetropolitan Areas, *Growth and Change*, vol. 19, no.1, pp. 13-30

Batten, D.F. and B. Johansson (1987), The Dynamics of Metropolitan Change, *Geographical Analysis*, vol. 19, nr.3, pp. 189-199

Batten, D.F. (1982), On the Dynamics of Industrial Evolution, *Regional Science and Urban Economics*, vol.12, pp. 449-462

Bearse, P.J. (1978), On the Intra-Regional Diffusion of Business Service Activity, *Regional Studies*, vol.12, pp. 563-578

Berry, B.J.L. (1972), Hierarchical Diffusion: The Basis of Developmental Filtering and Spread in a System of Growth Centers, in: N.M. Hansen (ed), *Growth Centers in Regional Economic Development*, pp. 108-138, The Free Press, New York

Birch, D.L. (1979), *The Job Generation Process*, Massachusetts Institute of Technology, Cambridge

Bishop, Y.M.M., S.E. Fienberg and P.W. Holland (1977), *Discrete Multivariate Analysis: Theory and Practice*, MIT Press, Massachusetts

Blommestein, H. and P. Nijkamp (1987), Adoption and Diffusion of Innovations and Evolution of Spatial Systems, in: D. Batten, J. Casti and B. Johansson (eds), *Economic Evolution and Structural Adjustment*, Springer-Verlag, Berlin, pp. 368-380.

Bluestone, B. and B. Harrison (1982), *Deindustrialization of America*, Basis Books, New York

Bouwman, H., T. Thuis and A. Verhoef (1985a), High Tech in Nederland. Vestigingsplaatsfactoren en Ruimtelijke Spreiding, R.U. Utrecht/Free University Amsterdam

Bouwman, H. and A. Verhoef (1985b), Ruimtelijke Spreiding van High Tech Werkgelegenheid, Free University, Amsterdam

Brown, L.A. (1981), *Innovation Diffusion. A New Perspective*, Methuen, London

Bureau of Statistics of Amsterdam, Amsterdam in Cijfers, several volumes, Afdeling Onderzoek en Statistiek, Bestuursinformatie Gemeente Amsterdam

Bushwell, R.J. and E.W. Lewis (1970), The Geographic Distribution of Industrial Research Activity in the U.K., *Regional Studies*, vol. 4, pp. 297-306

Bushwell, R.J. (1983), Research and Development: A Review, in: A. Gillespie (ed), *Technological Change and Regional Development*, pp. 9-22, Pion, London

Button, K. (1988), High-Technology Companies: An Examination of their Transport Needs, *Progress in Planning*, vol. 29, no. 2, pp. 79-146

Buunk, J.B. and P.C.M. Elderman (1982) De dynamiek van het bedrijvenbestand in Rijnmond, *Economisch Statistische Berichten*, pp. 1286-1290

Buursink, J. (1985), *De Dienstensector in Nederland. Een Geografisch Portret*, Van Gorcum, Assen

Camagni, R. (1984), Spatial Diffusion of Pervasive Process Innovation, Paper presented at the twenty-fourth European Congress of the Regional Science Association, Milan, August 28-31

Camagni, R. and R. Rabellotti (1986), Innovation and Territory: The Milan High-Tech and Innovation Field; Paper presented at the GREMI Seminar on "Les Regions et la Diffusion des Technologies Nouvellles", Paris, 1986

Camagni, R. and L. Diappi (1985), Urban Growth and Decline in a Hierarchical System: A Supply-Side Dynamic Approach, Paper presented at the IIASA Workshop on Dynamics of Metropolitan Areas, Rotterdam 1985

Carlino, G.A. (1978), *Economies of Scale in Manufacturing Location*, Martinus Nijhoff, Leiden

Centraal Bureau voor de Statistiek, *Nationale Rekeningen* (National Accounts), several volumes, CBS, Voorburg

Centraal Bureau voor de Statistiek, *Statistiek Werkzame Personen* (Statistics of Employed Persons), 1973-1985, CBS, Voorburg

Centraal Bureau voor de Statistiek, *Regionaal Statistisch Zakboek*, several volumes, CBS Voorburg

Centrale Databank Woerden (1987), Adressencatalogus 1987/88, Woerden

Christaller, W. (1933), *Die Zentralen Orte in Süddeutschland*, Jena

Clark, J., C. Freeman and L. Soete (1981a), Long Waves, Inventions and Innovations, *Futures*, vol. 13, pp. 308-22 (special issue)

Clark, J., C. Freeman and L. Soete (1981b), Long Waves and Technological Developments in the 20th Century, in: D. Petzina and G. van Roon (eds), *Konjunktur, Krise, Gesellschaft*, pp. 132-169, Klett-Cotta, Stuttgart

Coombs, R.W. and A. Kleinknecht (1984), New Evidence of the Shift towards Process Innovation during the Long Wave Upswing, in: C. Freeman (ed), *Design, Innovation and Long Cycles in Economic Development*, pp. 127-158, Pinter, London

Daniels, P.W. (1987), Internationalization of Producer Services and Metropolitan Development, paper presented at the conference on 'The Future of the Metropolitan Economy', The Hague, 1987

Daniels, P.W. (1985), *Service Industries*, Methuen, London

Davelaar, E.J. and P. Nijkamp (1986), De Stad als Broedplaats van Nieuwe Activiteiten, *Stedebouw en Volkshuisvesting*, no.2, pp. 61-66

Davelaar, E.J. and P. Nijkamp (1987a), The Urban Incubator Hypothesis: Old Wine in New Bottles? in: M.M. Fischer and M. Sauberer (eds) Gesellschaft-Wirtschaft-Raum, pp. 198-213, AMR-Info, vol.17, Melzer, Wien

Davelaar, E.J. and P. Nijkamp (1987b), De Incubatie-Hypothese: Stedelijk Reveil door middel van Innovatie?, Maandschrift Economie, vol. 51, no. 4, pp. 250-262

Davelaar, E.J. and P. Nijkamp (1987c), Het Grootstedelijk Milieu als Informatie- en Incubatiepool voor Innovaties, Planning, vol. 30, pp. 716-722

Davelaar, E.J. and P. Nijkamp (1987d), Industriële Innovaties en de Broedplaatsgedachte, Economisch Statistische Berichten, vol.72, no.3617, pp. 716-722

Davelaar, E.J. and P. Nijkamp (1987e), Technologiebeleid en Regiobeleid. Een Nederlandse Verkenning, Planologisch Nieuws, vol. 7, no. 4, pp. 190-195

Davelaar, E.J. and P. Nijkamp (1988), The Urban Incubator Hypothesis: Re-vitalization of Metropolitan Areas? in: The Annals of Regional Science, vol. 22, no. 3, pp. 48-65 (special issue)

Davelaar, E.J. and P. Nijkamp (1989a), Technological Innovation and Spatial Transformation, in: H.J. Ewers (ed), Technological Innovation and Regional Development Policy, De Gruijter, Berlin (forthcoming)

Davelaar, E.J. and P. Nijkamp (1989b), New Technology Systems in Space: The Case of Producer Services, in: D. Maillat (ed), Revue d'Economie Régionale et Urbaine (forthcoming)

Davelaar, E.J. and P. Nijkamp (1989c), Technological Innovation and Spatial Transformation, Technological Forecasting and Social Change, (forthcoming)

Davelaar, E.J. and P. Nijkamp (1989d), Spatial Dispersion of Technological Innovation. A Case Study for the Netherlands by means of Partial Least Squares, Journal of Regional Science (forthcoming)

Davelaar, E.J. and P. Nijkamp (1989e), Innovative Behavior of Industrial Firms: Results from a Dutch Empirical Study, in: A. Andersson, D.F. Batten and C. Karlsson (eds), Knowledge and Industrial Organization, Springer Verlag, Berlin (forthcoming)

Davelaar E.J. and P. Nijkamp (1989f), Structural Transformation of Cities: The Case of the Netherlands, in: H.J. Ewers and P. Nijkamp (eds), Urban Sustainability, Gower, Aldershot (forthcoming)

Davelaar, E.J. and P. Nijkamp (1989g), Spatial Dispersion of Technological Innovation: The Incubator Hypothesis, to be published in the proceedings of the conference on 'Innovation Diffusion', Venice, March 1986

327

Davelaar. E.J. and P. Nijkamp (1989h), The Role of the Metropolitan Milieu as an Incubation Centre for Technological Innovations A Dutch Case Study, *Urban Studies* (forthcoming)

Davies, S. (1979), *The Diffusion of Process Innovations*, Cambridge University Press, Cambridge

Dinteren, J.H.J. van and E.C.J. Kraan (1988), De Betekenis van Zakelijke Diensten, *Economisch Statistische Berichten*, vol. 73, pp.431-435

Dosi, G. (1984), *Technical Change and Industrial Transformation*, Mac Millan, Hong Kong

Dosi, G., C. Freeman, R. Nelson, G. Silverberg and L. Soete (eds) (1989), *Technical Change and Economic Theory*, Pinter, London

Duijn, J.J. van (1981), Fluctuations in Innovations over Time, *Futures* vol. 13, no. 4, pp. 264-275 (special issue)

Duijn, J.J. van (1983), *The Long Wave in Economic Life*, Allen & Unwin, London

EIM (1986), Structuur en Ontwikkeling van de Commerciële Dienstensector in Nederland, EIM, Zoetermeer

Ewers, H.J. (1986) Spatial Dynamics of Technological Developments and Employment Effects, in: P. Nijkamp (ed), *Technological Change, Employment and Spatial Dynamics*, pp. 157-176, Springer-Verlag, Berlin

Ewers, H.J., R.W. Wettman, F.J. Bade, J. Kleine and H. Kirst (1979), Innovationsorientierte Regionalpolitik, Research Report, Wissenschaftszentrum, Berlin

Ewers, H.J. and R.W. Wettman (1980), Innovation-oriented Regional Policy, *Regional Studies*, vol. 14, pp. 161-179

Ewers, H.J., and P. Nijkamp (1989) (eds), *Urban Sustainability*, Gower, Aldershot (forthcoming)

Feller, I. (1973), Determinants of the Composition of Urban Inventions, *Economic Geography*, vol. 49, pp. 47-58

Feller, I. (1979), Three Coigns on Diffusion Research, *Knowledge: Creation, Diffusion, Utilization*, vol. 1, pp. 293-312

Folmer, H. (1985), *Regional Economic Policy*, Martinus Nijhoff, Dordrecht

Fornell, C. (ed) (1982), *A Second Generation of Multivariate Analysis*, vols 1 and 2, Praeger, New York

Freeman, C. (1987a), Technical Innovation, Long Cycles and Regional Policy, in: K. Chapman and G. Humphrys (eds), *Technical Change and Industrial Policy*, pp. 10-25, Basic Blackwell, Oxford

Freeman, C. (1987b), Technical Innovation, Diffusion, and Long Cycles of Economic Development, in: T.Vasko (ed) *The Long-Wave Debate*, pp. 295-309, Springer Verlag, Berlin

Freeman, C., R.C. Curnow, J.K. Fuller, A.B. Robertson, P.J. Whittaker (1968), Chemical Process Plant: Innovation and the World Market, National Institute Economic Review, no. 45

Freeman, C., J. Clark and L. Soete (1982), *Unemployment and Technical Innovation. A Study of Long Waves and Economic Development*, Frances Pinter, London

Gellman Research Associates (1976), Indicators of International Trends in Technological Innovation, National Science Foundation, Washington

Giaoutzi, M., P. Nijkamp and D. Storey (1988), *Small and Medium-Size Firms and Regional Development*, Croom Helm, London

Gibbs, D.C. and A.T. Thwaites (1985), The Location and Potential Mobility of Research and Development Activity : A Regional Perspective; Paper presented at the twenty-fifth European congress of the Regional Science Association

Granstrand, O. (1986), The Modelling of Buyer/Seller-Diffusion Processes. A Novel Approach to Modelling Diffusion and Simple Evolution of Market Structure, paper presented at the conference on 'Innovation Diffusion', Venice, March 1986

Griliches, Z. (1957), Hybrid Corn: An Exploration in the Economics of Technological Change, *Econometrica*, vol. 25, pp. 501-522

Gudgin, G. (1978), *Industrial Location Processes and Regional Employment Growth*, Saxon House, Farnborough

Haan, M.A. de and P. Tordoir (1986), Het Belang van Diensten voor de Nederlandse Economie. Gangbare en Nieuwe Inzichten, PSC-TNO, Delft

Hägerstrand, T. (1967), *Innovation Diffusion as a Spatial Process*, University of Chicago Press, Chicago

Hall, P. (1984), *The World Cities*, Weidenfeld and Nicholson, London

Hansen, E.R. (1983), Why Do Firms Locate Where They Do? World Bank Discussion Paper, Wudd 25

Hansen, J.A. (1986), Innovation Characteristics of Industries in the United States, paper presented at the conference 'Technologie, Arbeid en Economie', Maastricht 23-24 October 1986

Hekman (1980), J.S. (1980), The Product Cycle and New England Textiles, *Quarterly Journal of Economics*, vol.94, pp. 697-717

Hirsch, S. (1967), *Location of Industry and International Competitiveness*, Clarendon Press, Oxford

Hirschman, A.O. (1958), *The Strategy of Economic Development*, Yale University Press, New Haven.

Hoover, E.M. and R. Vernon (1959), *Anatomy of a Metropolis*, Harvard University Press, Cambridge

Howells, J.R.L. (1984), The Location of Research and Development: Some Observations and Evidence from Britain, *Regional Studies*, vol. 18, pp. 13-29

Huppes, T. (1987), Over Netwerken en Organisaties, *Economisch Statistische Berichten*, pp. 932-940

Jacobs, J. (1966), *The Death and Life of Great American Cities*, Vintage Books, London

Jansen, A.C.M. and M. de Smidt (1974), *Industrie en Ruimte*, Van Gorcum, Assen

Jobse, R.B. and B. Needham (1988), The Economic Future of the Randstad Holland, *Urban Studies*, vol.25, pp. 282-296

Johansson, B. and J. Larsson (1989), Innovation Processes in the Urban Network of Export and Import Nodes: A Swedish Example, in: H.J. Ewers and P. Nijkamp (eds), *Urban Sustainability*, Gower, Aldershot (forthcoming)

Johansson, B. and P. Nijkamp (1987), Analysis of Episodes in Urban Event Histories, in: L. van den Berg, L.S. Burns and L.H. Klaassen (eds), *Spatial Cycles*, pp. 43-66, Gower Aldershot

Jong, M. de and H.M. Paap (1983), De Stedelijke Op- en Neergang van Amsterdam, Department of Economics, Free University, Amsterdam

Jong, H.W. de (1985), *Dynamische Markttheorie*, Stenfert Kroese, Leiden

Jong, M.W. de and J.G. Lambooy (1984), De Informatica-Sector Centraal; Perspectieven voor de Amsterdamse Binnenstad, EGI University of Amsterdam, Amsterdam

Jong, M.W. de (1984), Ruimtelijke Dynamiek van het Midden- en Kleinbedrijf; het Lokatiepatroon van Kleine Industriële Ondernemingen in de Noordvleugel van de Randstad, EGI University of Amsterdam, Amsterdam

Jöreskog, K.G. and H. Wold (eds) (1982), *Systems under Indirect Observation: Causality, Structure, Prediction*, 2 vols., North-Holland, Amsterdam

Kamann, D.J.F. and P. Nijkamp (1988), Technogenesis: Incubation and Diffusion, paper presented at the European Summer Institue on 'Theories and Policies of Technological Development at the Local Level', Arco, Italy, 17-23 July 1988

Kamien, M.I. and N.L. Schwartz (1982), *Market Structure and Innovation*, Cambridge University Press, Cambridge

Karlsson, C. (1988), Innovation Adoption and the Product Life Cycle, Umeå Economic Studies no. 185, University of Umeå

Keeble, D., P.L. Owens and C. Thompson (1983), The Urban-Rural Manufacturing Shift in the European Community, *Urban Studies*, vol. 20. pp. 405-418

Keeble, D. (1986), The Changing Spatial Structure in the United Kingdom in: H.J. Ewers, J.B. Goddard and H. Matzerath (eds), *The Future of the Metropolis*, pp. 171-199, de Gruyter, Berlin

Keeble, D. and T. Kelly (1986), New Firms and High-Technology Industry in the United Kingdom: the Case of Computer Electronics, in: D. Keeble and E. Wever (eds) *New Firms and Regional Development in Europe*, pp. 75-104, Croom Helm, London

Kendrick, J. (1961) *Productivity Trends in the United States*, Princeton University Press, Princeton

Kleinknecht, A. (1981), Observations on the Schumpeterian Swarming of Innovations, *Futures*, vol. 13, pp. 293-307 (special issue)

Kleinknecht, A. (1987a), *Innovation Patterns in Crisis and Prosperity*, MacMillan, London

Kleinknecht, A. (1987b), *Industriële Innovatie in Nederland. Een Enquête-Onderzoek*, Van Gorcum, Assen

Kleinknecht, A. (1987c), Rates of Innovations and Profits in the Long Wave, in: *The Long-Wave Debate*, T. Vasko (ed), pp. 216-238, Springer Verlag, Berlin

Kleinknecht, A. and B. Verspagen (1987), R&D en Concurrentie-vermogen, in: *Economisch Statistische Berichten*, pp. 343-345

Kleinknecht, A. and A. Mouwen (1985), Regional Innovatie (R en D): Verschuiving naar de 'Halfwegzone'? in: W.T.M. Molle (ed) *Innovatie en Regio*, pp. 125-142, Staatsuitgeverij, Den Haag

Knaap, G.A. van der and W.F. Sleegers (1980), De Structuur van Migratiestromen in Nederland, Report nr. 1, Serie C - nr. 80-4, Economic Geographic Institute, Erasmus University Rotterdam

Knaap, G.A. van der and E. Wever (eds) (1987), *New Technology and Regional Development*, Croom Helm, London

Knaap, G.A. van der and M.S. van Geenhuizen (1988), A Longitudinal Analysis of the Growth of Firms, Economisch Geografisch Instituut, Discussie-Paper no. 88-9, Erasmus University Rotterdam

Koerhuis, H. and W. Cnossen (1982), De Software- en Computerservice-bedrijven, Sociaal Geografische Reeks, no. 23, GIRUG, Groningen

Kok, J.A.A.M., G.J.D. Offerman and P.H. Pellenbarg (1985), Innovatieve Bedrijven in het Grootstedelijk Milieu, Report Series Social Geography, nr. 34-1985, Geographic Institute, State University, Groningen

Krumme, G. and R. Hayter (1975), Implications of Corporate Strategies and Product Cycle Adjustment for Regional Employment Changes, in: L. Collins and D.F. Walker, *Locational Dynamics of Manufacturing Activity*, pp. 325-356, London

Kuznets, S. (1930), *Secular Movements in Production and Prices*, Houghton Mifflin, Boston

Kuznets, S. (1940), Schumpeter's Business Cycles, *American Economic Review*, vol. 30, pp. 257-271

Lambooy, J.G. and N. Tates (1983), Zakelijke Diensten, een complementaire Sector, *Economisch Statistische Berichten*, pp. 676-680

Lambooy, J.G. (1978), Bedrijvigheid en het Grootstedelijk Productiemilieu: Kansen en Bedreigingen, in: N.A. de Boer and W.F. Heinemeyer (eds), *Het Grootstedelijk Milieu*, pp. 19-40, Van Gorcum, Assen

Lambooy, J.G. (ed) (1984), *New Spatial Dynamics and Economic Crisis*, Finnpublishers, Tampere

Lambooy, J.G. and P. Tordoir (1985), Professional Services and Regional Development, Fast-paper no.74, European Commission, Brussel

Leigh, R., D. North and L. Steinberg (1986), Restructuring and Locational Change in London's Electronics Industries, in: H.J. Ewers, J.B. Goddard and H. Matzerath (eds), *The Future of the Metropolis*, pp. 251-284, de Gruyter, Berlin

Leone, R.A. and R. Struyck (1976), The Incubator Hypothesis: Evidence from five SMSAs, *Urban Studies*, vol.13, pp. 325-331

Lohmöller, J.B. (1984), LVPLS Program Manual, University of Cologne, Cologne

Louter, P.J. (1987), Regionaal-Economische Vernieuwing in Nederland, EGI-publicatie nr. 87-1, Erasmus University, Rotterdam

Machielse, C. and P.A. de Ruijter (1988), Economisch-Technologische Vernieuwing en Ruimtelijke Organisatie, INRO-TNO, Delft

Mahdavi, K.B. (1972), *Technological Innovation: An Efficiency Investigation*, Beckmans, Stockholm

Malecki, E.J. (1979a), Agglomeration and Intra-Firm Linkage in R&D Location in the United States, *TESG*, vol. 70, pp. 322-331

Malecki, E.J. (1979b), Locational Trends in R&D by large U.S. Corporations 1965-1977, *Economic Geography*, vol.55, pp. 309-323

Malecki, E.J. (1980), Corporate Organization of R and D and the Location of Technological Activities, *Regional Studies*, vol 14, pp. 219-234

Malecki, E.J. (1983), Technology and Regional Development: A Survey, *International Regional Science Review*, vol. 8, pp. 89-125

Malecki, E.J. and P. Varaiya (1986), Innovation and Changes in Regional Structure, in: P. Nijkamp (ed), *Handbook of Regional and Urban Economics*, pp. 629-645, vol.7, North-Holland, Amsterdam

Malecki, E.J. and P. Nijkamp (1988), Technology and Regional Development: Some Thoughts on Policy, *Environment and Planning C: Government and Policy*, vol. 6, pp. 383-399

332

Mansfield, E. (1968), *Industrial Research and Technological Innovation*. Norton, New York

Markusen, A., P. Hall and A. Glasmeier (1986), *High Tech America*, Allen & Unwin, London

Marshall, N. (1985), Business Services, the Regions and Regional Policy, *Regional Studies*, vol. 19, pp. 533-563

Martin, F. N. Swan, I. Banks, G. Barker and R. Beaudry (1979), The Interregional Diffusion of Innovations in Canada, Ministry of Supply and Services, Canada

May, R.M. (1976), Simple Mathematical Models with very Complicated Dynamics, *Nature*, vol.261, pp. 459-467.

McArthur, R. (1987), Innovation, Diffusion and Technical Change: A Case Study, in : K. Chapman and G. Humphrys (eds), *Technical Change and Industrial Policy*, pp. 26-50, Basic Blackwell, Oxford

Meer, H.J. van der and M. Brand (1987), Innovatiepotentie van Steden, *Economisch Statistische Berichten*, vol. 3611, pp. 600-602

Meer, J.D. van der and J.J. van Tilburg (1983), Spin-offs uit de Nederlandse Kenniscentra, Projekt Technologiebeleid/ Ministerie van Economische Zaken

Mensch, G. (1981a), *Stalemate in Technology: Innovations Overcome the Depression*, Ballinger, New York

Mensch, G. (1981b), Long Waves and Technological Developments in the 20th Century; Comment in: *Konjunktur, Krise, Gesellschaft*, D. Petzina and G. van Roon (eds), pp. 170-179, Klett-Cotta, Stuttgart

Mensch, G., C. Coutinho and K. Kaasch (1981), Changing Capital Values and the Propensity to Innovate, in: *Futures*, vol. 13, pp. 276-292 (special issue)

Metcalfe, J. (1981), Impulse and Diffusion in the Study of Technical Change, *Futures*, vol. 13, no.5, pp. 347-359 (special issue)

Molle, W.T.M. and J.G. Vianen (1982), Het Vertrek van Bedrijven uit Amsterdam, *Economisch Statistische Berichten*, pp. 608-613

Molle, W.T.M. and J.G. Vianen (1985), De Vestigingsplaats van enkele Innovatieve Bedrijfsgroepen, in: W.T.M. Molle (ed) *Innovatie en Regio*, pp. 77-100, Staatsuitgeverij, Den Haag

Molle, W.T.M. (ed) (1985), *Innovatie en Regio*, Staatsuitgeverij, Den Haag

Moore, B. and R. Spires (1983), The Experience of the Cambridge Science Park, paper presented at the workshop 'Research, Technology and Regional Policy, Paris, 24-27 October 1983

Morphet, C.S. (1987), Research, Development and Innovation in the Segmented Economy: Spatial Implications, in: G.A. van der Knaap and E. Wever (eds), *New Technology and Regional Development*, pp. 45-62, Croom Helm, London

Moss, M.L. (1985), Telecommunications and the Future of Cities, paper presented at the Conference on Landtronics, London

Mouwen, A. (1984), Theorie en Praktijk van Lange-Termijn Stedelijke Ontwikkelingen (een case-study voor Amsterdam), discussienota 1984-7, Department of Economics, Free University, Amsterdam

Mouwen, A. and P. Nijkamp (1985), Knowledge Centres as Strategic Tools in Regional Policy, Research Memorandum, Department of Economics, Free University, Amsterdam

Nabseth, L and G.F. Ray (eds) (1974), *The Diffusion of New Industrial Processes: An International Study*, Cambridge University Press, Cambridge

Naisbitt (1984), *Megatrends. Ten New Directions Transforming Our Lives*, Warner Books, New York

Nelson, R. (1986), The Generation and Utilization of Technology: A Cross Industry Analysis; paper presented at the conference on "Innovation Diffusion", March 1986, Venice

Nelson, R. and S.G. Winter (1977), In Search of a Useful Theory of Innovation, *Research Policy*, vol. 6, pp. 36-76

Nelson, R.R. and S.G. Winter (1982), *An Evolutionary Theory of Economic Changes*, Harvard University Press, Cambridge, Mass.

Netherlands Economic Institute (1984), *Technologische Vernieuwing en Regionale Ontwikkeling in Nederland (Transfer)*, NEI, Rotterdam

Nijkamp, P. (1983), Technological Change, Policy Response and Spatial Dynamics, in: D.A. Griffith and A.C. Lea (eds), *Evolving Geographical Structures*, pp. 75-98, Martinus Nijhoff, The Hague

Nijkamp, P. and U. Schubert (1983), Structural Change in Urban Systems, Contributions to the Metropolitan Study, no.5, IIASA, Laxenburg

Nijkamp, P., H. Leitner and N. Wrigley (eds) (1985), *Measuring the Unmeasurable*, Martinus Nijhoff, Dordrecht

Nijkamp, P. (1986), (ed), *Technological Change, Employment and Spatial Dynamics*, Springer Verlag, Berlin.

Nijkamp, P. (1987), New Technology and Regional Development, in: T. Vasko (ed), *The Long-Wave Debate*, pp. 274-282, Springer-Verlag, Berlin

Nijkamp, P. (1988), Information Center Policy in a Spatial Development Perspective, *Economic Development and Cultural Change*, pp. 173-193, vol. 37, no. 1

Northcott, J. and P. Rogers (1984), Microelectronics in British Industry: The Pattern of Change, Policy Studies Institute, London

334

Norton, R.D. (1979), *City Life Cycles and American Urban Policy* Academic Press, New York

Norton, R.D. and J. Rees (1979), The Product Cycle and the Spatia Decentralization of American Manufacturing, *Regional Studies*, vol. 13 pp. 141-151

Noyelle, T.J. and T.M. Stanback (1984), *The Economic Transformation o American Cities*, Rowman & Allanheld, Totowa, N.J.

Nozeman, E.F. (1980), Tussen Handwerkslieden en Beleidsadviseurs Honderd Jaar Bedrijvigheid in Amsterdam, in: *Wonen en Werken en Verkee. in Amsterdam 1880-1980*, pp. 103-178, Geografisch en Planologiscl Instituut, Free University, Amsterdam.

Oakey, R.P., A.T. Thwaites and P.A. Nash (1980), The Regiona. Distribution of Innovative Manufacturing Establishments in Britain *Regional Studies*, vol. 14, pp. 235-253

Oakey, R.P. (1984), Innovation and Regional Growth in Small Higl Technology Firms: Evidence from Britain and the USA, *Regional Studies* vol. 18, pp. 237-251

Orishimo, I., G.J.D. Hewings and P. Nijkamp (eds) (1988), *Informatior Technology; Economic and Spatial Dimensions*, Springer Verlag, Berlin

OTA (1984), *Technology, Innovation, and Regional Economic Development.* Office of Technology Assessment, Washington D.C.

Pedersen, P.O. (1975), *Urban-Regional Development in South America: . Process of Diffusion and Integration*, Mouton, The Hague

Pedersen, P.O. (1970), Innovation Diffusion within and between Nationa] Urban Systems, *Geographical Analysis*, vol.2, pp. 203-254

Perez, C. (1983), Structural Change and the Assimilation of Nev Technologies in the Economic and Social System; in *Futures*, vol. 15, pp. 357-375

Porter, M.E. (1980), *Competitive Strategy*, MacMillan, London

Potters, A.L.M. (1983), Dienstverlening in beweging, *Stedebouw er Volkshuisvesting*, pp. 169-176

Pred, A.R. (1977), *City-Systems in Advanced Economies. Past Growth, Present Processes and Future Development Options*, Hutchinson, London

Premus, R. (1982), Location of High Technology Firms and Regional Development, Joint Study Economic Committee, U.S. Congress, Washington D.C.

Roobeek, A.J.M. (1988), Een Race zonder Finish. De Rol van de Overheid in de Technologiewedloop, Free University Press, Amsterdam

Rosenberg, N. (1976), *Perspectives on Technology*, Cambridge University Press, Cambridge, Mass.

Rothwell, R. and W. Zegveld (1985), *Reindustrialization and Technology*, Longman, Essex

Ruijter, P.A. de (1983), De Bruikbaarheid van het Begrip 'Incubatiemilieu', *KNAG Geografisch Tijdschrift*, vol. 17, nr 2, pp. 106-110

Ruijter, P.A. de, M. Olij and A.J.M. Leijten (1986), Kansrijk in Profiel Lokatiepatronen en functioneren van Kansrijke Bedrijvigheid, TNO/PSC, Delft Apeldoorn

Sahal, D. (1980), The Network and Significance of Technological Cycles, *International Journal of Systems Science*, vol. 11, no.8, pp. 985-1000

Sahal, D. (1981), Alternative Conceptions of Technology, *Research Policy*, vol. 10, pp. 3-24

Sauer, C.O. (1952), *Agricultural Origins and Dispersals*, American Geographical Society, New York

Schmookler, J. (1966), *Invention and Economic Growth*, Harvard University Press, Cambridge, Mass.

Schumpeter, J.A. (1934), *The Theory of Economic Development*, Harvard University Press, Cambridge, Mass.

Schumpeter, J.A. (1943), *Capitalism, Socialism and Democracy*, Harper and Row, New York

Schumpeter, J.A. (1939), *Business Cycles: A Historical and Statistical Analysis of the Capitalist Process*, Mc Graw-Hill, New York

Schwamp, E.W. (1987), Technology Parks and Interregional Competition in the Federal Republic of Germany, in G.A. van der Knaap and E. Wever (eds), *New Technology and Regional Development*, pp. 119-135, Croom Helm, London

Scott, A.J. (1982), Locational Patterns and Dynamics of Industrial Activitiy in the Modern Metropolis: a Review Essay, *Urban Studies*, pp. 111-142

SKIM (1981), Een Microscoop voor Economen, SKIM, Rotterdam

Smidt, M. de, J.C. van Opheusden and H.T. Rimmelzwaan (1987), *Teleshopping in Ruimtelijk Perspectief*, STOGO, Utrecht

Smidt, M. de and M. Conijn (1989), Elektronisch Winkelen, in: M. Bierman and M.J. Dijst (eds), *Tele-Netwerken en Systemen. Actuele Balans van Onderzoek Beleid en Praktijk*, pp. 93-101, SISWO-publicatie 337, Amsterdam

Stöhr, W.B. (1985), Territorial Innovation Complexes, Paper (draft) of the Interdisciplinary Institute for Urban and Regional Studies, Vienna

Stokman, C.T.M. (1985), Jong en Klein in de Industrie: Ontwikkelingen en Mogelijkheden, Paper presented at the Symposium Technologie-Economie, September 1985, The Hague (mimeographed)

Stoneman, P. (1983), *The Economic Analysis of Technological Change*, Oxford University Press, Oxford

Storper, M. (1986), Technology and New Regional Growth Complexes: The Economics of Discontinuous Spatial Development, in: P. Nijkamp (ed) *Technological Change, Employment and Spatial Dynamics*, pp. 46-75 Springer-Verlag, Berlin

Taylor, M. (1987), Enterprise and the Product-Cycle Model: Conceptual Ambiguities, in: G.A. van der Knaap and E. Wever (eds), *New Technology and Regional Development*, pp. 75-93, Croom Helm, London

Terleckyj, N. (1974), *The Effects of R & D on Productivity Growth in Industry*, National Planning Association, Washington

Thomas, M.D. (1987), The Innovation Factor in the Process of Microeconomic Industrial Change: Conceptual Explorations, in: G.A. van der Knaap and E. Wever (eds), *New Technology and Regional Development*, pp. 21-44, Croom Helm, London

Thompson, W.R. (1968), Internal and External Factors in the Development of Urban Economies, in: H.S. Perloff and L. Wingo (eds), *Issues in Urban Economics*, pp. 43-62, The John Hopkins Press, Baltimore

Thorngren, B. (1970), How do Contact Systems affect Regional Development, *Environment and Planning*, vol.2, pp. 409-427

Thwaites, A.T. (1978), Technological Change, Mobile Plants and Regional Development, *Regional Studies*, vol. 12, pp. 445-461

Thwaites, A.T. (1982), Some Evidence of Regional Variations in the Introduction and Diffusion of Industrial Products and Processes Within British Manufacturing Industry, *Regional Studies*, vol. 16, pp. 371-381

Tuyl, J.M.C. (1987), R&D en Concurrentie-Vermogen, *Economisch Statistische Berichten*, pp. 140-143

Vasko, T. (1987), *The Long-Wave Debate*, Springer Verlag, Berlin

Vernon, R. (1960), *Metropolis 1985*, Harvard University Press, Cambridge

Vernon, R. (1966), International Investment and Institutional Trade in the Product Cycle, *Quarterly Journal of Economics*, vol.80, pp. 190-207

Vlessert, H.H. and C.P.A. Bartels (1985), *Kenniscentra als Elementen van het Regionaal Produktiemilieu*, Buro Bartels, Oudemolen

Wever, E. (1984), *Nieuwe ondernemingen in Nederland*, Van Gorcum, Assen

Wever, E. (1985), Regionaal Economisch Perspectief, Katholieke Universiteit Nijmegen

Winter, S.G. (1984), Schumpeterian Competition in Alternative Technological Regimes, *Journal of Economic Behavior and Organization*, vol. 5, pp. 287-320

Wold, H. (1985b), Systems Analysis by Partial Least Squares, in: P. Nijkamp, H. Leitner and N. Wrigley (eds), *Measuring the Unmeasurable*, pp. 221-249, Martinus Nijhoff, Dordrecht

Wold, H. (1985a), Partial Least Squares, in: J. Kotz-Johnson (ed), *Encyclopedia of Statistical Sciences*, vol.6, pp. 581-591

Wold, H. (1982), Soft Modeling: The Basic Design and Some Extensions, in: K.G. Jöreskog and H. Wold (eds), *Systems under Indirect Observation: Causality, Structure, Prediction*, vol. 2, pp. 1-54, North-Holland, Amsterdam

Wolff, S. de (1921), Prosperiteits- en Depressieperioden, *De Socialistsche Gids; Maandschrift der Sociaal Democratische Arbeiderspartij*, vol. VI, no. 1

Zelinski, W. (1967), Classical Town Names in the United States, *Geographical Review*, vol. 57, pp. 463-495

Zwan, A. van der (1979), On the Assesment of the Kondratiev Cycle and related Issues, Centre for Research in Business Economics, Erasmus University Rotterdam (mimeo)